The Green
and Virtual
Data Center

The Green and Virtual Data Center

Greg Schulz

CRC Press
Taylor & Francis Group
Boca Raton London New York

CRC Press is an imprint of the
Taylor & Francis Group, an **informa** business

CRC Press
Taylor & Francis Group
6000 Broken Sound Parkway NW, Suite 300
Boca Raton, FL 33487-2742

First issued in paperback 2019

© 2009 by Taylor & Francis Group, LLC
CRC Press is an imprint of Taylor & Francis Group, an Informa business

No claim to original U.S. Government works

ISBN-13: 978-1-4200-8666-9 (hbk)
ISBN-13: 978-0-367-38600-9 (pbk)

**Visit the Taylor & Francis Web site at
http://www.taylorandfrancis.com**

**and the Auerbach Web site at
http://www.auerbach-publications.com**

Contents

Preface   xiii

About the Author   xvii

Acknowledgments   xix

PART I: Green IT and the Green Gap-Real or Virtual?   1

Chapter 1   IT Data Center Economic and Ecological Sustainment   3

    1.1   The Many Faces of Green—Environmental and
         Economic   3
    1.2   The Growing Green Gap: Misdirected Messaging,
         Opportunities for Action   5
    1.3   IT Data Center "Green" Myths and Realties   7
    1.4   PCFE Trends, Issues, Drivers, and Related Factors   10
    1.5   Closing the Green Gap for IT Data Centers   15
        1.5.1   Energy Consumption and Emissions:
            Green Spotlight Focus   20
        1.5.2   EHS and Recycling: The Other
            Green Focus   21
        1.5.3   Establishing a Green PCFE Strategy   22
    1.6   Summary   27

**Chapter 2   Energy-Efficient and Ecologically Friendly
           Data Centers**   29

    2.1   Electric Power and Cooling Challenges   30
    2.2   Electrical Power—Supply and Demand Distribution   33
    2.3   Determining Your Energy Usage   37
    2.4   From Energy Avoidance to Efficiency   39

2.5   Energy Efficiency Incentives, Rebates, and
        Alternative Energy Sources   41
2.6   PCFE and Environmental Health and
        Safety Standards   45
2.7   Summary   45

PART II: Next-Generation Virtual Data centers   **47**

**Chapter 3   What Defines a Next-Generation and Virtual
        Data Center?**   **49**

3.1   Why Virtualize a Data Center?   50
3.2   Virtualization Beyond Consolidation—Enabling
        Transparency   54
3.3   Components of a Virtual Data Center   56
        3.3.1   Infrastructure Resource Management
                  Software Tools   59
        3.3.2   Measurements and Management Insight   59
        3.3.3   Facilities and Habitats for Technology   60
        3.3.4   Tiered Servers and Software   61
        3.3.5   Tiered Storage and Storage Management   62
        3.3.6   Tiered Networks and I/O Virtualization   62
        3.3.7   Virtual Offices, Desktops,
                  and Workstations   63
3.4   Summary   63

Chapter 4   IT Infrastructure Resource Management   **65**

4.1   Common IRM Activities   67
4.2   Data Security (Logical and Physical)   69
4.3   Data Protection and Availability for Virtual
        Environments   70
        4.3.1   Time to Re-Architect and Upgrade
                  Data Protection   71
        4.3.2   Technologies and Techniques—Virtual Server
                  Data Protection Options   74
        4.3.3   Virtual Machine Movement
                  and Migration   75
        4.3.4   High Availability   76
        4.3.5   Snapshots   77
        4.3.6   Agent-Based and Agent-Less
                  Data Protection   78
        4.3.7   Proxy-Based Backup   79

	4.3.8	Local and Remote Data Replication	81
	4.3.9	Archiving and Data Preservation	82
	4.3.10	Complete Data Protection	83
4.4	Data Protection Management and Event Correlation		84
4.5	Server, Storage, and Network Resource Management		86
	4.5.1	Search and eDiscovery	87
	4.5.2	Rescuing Stranded or Orphaned Resources	88
	4.5.3	Capacity, Availability, and Performance Planning	89
	4.5.4	Energy Efficiency and PCFE Management Software	91
4.6	Summary		92

Chapter 5   Measurement, Metrics, and Management of IT Resources   93

5.1	Data Center-Related Metrics	96
5.2	Different Metrics for Different Audiences	100
5.3	Measuring Performance and Active Resource Usage	107
5.4	Measuring Capacity and Idle Resource Usage	113
5.5	Measuring Availability, Reliability, and Serviceability	115
5.6	Applying Various Metrics and Measurements	116
5.7	Sources for Metrics, Benchmarks, and Simulation Tools	117
5.8	Summary	118

PART III: Technologies for Enabling Green and Virtual Data Centers   121

Chapter 6   Highly Effective Data Center Facilities and Habitats for Technology   123

6.1	Data Center Challenges and Issues	124
6.2	What Makes up a Data Center	129
	6.2.1	Tiered Data Centers	130
6.3	Data Center Electrical Power and Energy Management	132
	6.3.1	Secondary and Standby Power	134

6.3.2   Alternative Energy Options and
DC Power   136
6.4   Cooling, HVAC, Smoke and Fire Suppression   138
6.4.1   Cooling and HVAC   138
6.4.2   Physical Security   143
6.4.3   Smoke and Fire Detection
and Suppression   144
6.4.4   Cabinets and Equipment Racks   149
6.4.5   Environmental Health and
Safety Management   150
6.5   Data Center Location   151
6.6   Virtual Data Centers Today and Tomorrow   152
6.7   Cloud Computing, Out-Sourced, and
Managed Services   155
6.8   Data Center Tips and Actions   158
6.9   Summary   160

Chapter 7   Servers—Physical, Virtual, and Software   163

7.1   Server Issues and Challenges   164
7.2   Fundamentals of Physical Servers   172
7.2.1   Central Processing Units   175
7.2.2   Memory   178
7.2.3   I/O Connectivity for Attaching
Peripheral Devices   181
7.2.4   Cabinets, Racks, and Power Supplies   182
7.2.5   Measuring and Comparing Server
Performance   183
7.3   Types, Categories, and Tiers Of Servers   183
7.3.1   Blade Servers and Blade Centers   184
7.3.2   Virtual Servers   187
7.4   Clusters and Grids   200
7.5   Summary   201

Chapter 8   Data Storage—Disk, Tape, Optical, and Memory   205

8.1   Data Storage Trends, Challenges, and Issues   206
8.2   Addressing PCFE Storage Issues   209
8.3   Data Life Cycle and Access Patterns   210
8.4   Tiered Storage—Balancing Application Service
with PCFE Requirements   212
8.4.1   Tiered Storage System Architectures   213
8.4.2   Tiered Storage Media or Devices   218

	8.4.3	Intelligent Power Management and MAID 2.0	224
	8.4.4	Balancing PACE to Address PCFE Issues with Tiered Storage	226
8.5	Data and Storage Security		228
8.6	Data Footprint Reduction—Techniques and Best Practices		229
	8.6.1	Archiving for Compliance and General Data Retention	230
	8.6.2	Data Compression (Real-Time and Offline)	231
	8.6.3	De-duplication	232
	8.6.4	Hybrid Data Footprint Reduction— Compression and De-duplication	234
8.7	Countering Underutilized Storage Capacity		234
	8.7.1	Thin Provision, Space-Saving Clones	236
	8.7.2	How RAID Affect PCFE and PACE	236
8.8	Storage Virtualization—Aggregate, Emulate, Migrate		240
	8.8.1	Volume Mangers and Global Name Spaces	240
	8.8.2	Virtualization and Storage Services	241
8.9	Comparing Storage Energy Efficiency and Effectiveness		244
8.10	Benchmarking		246
8.11	Summary		247

Chapter 9	**Networking with Your Servers and Storage**		**249**
9.1	I/O and Networking Demands And Challenges		250
9.2	Fundamentals and Components		253
9.3	Tiered Access for Servers and Storage—Local and Remote		255
	9.3.1	Peripheral Component Interconnect (PCI)	256
	9.3.2	Local Area Networking, Storage, and Peripheral I/O	258
	9.3.3	Ethernet	260
	9.3.4	Fibre Channel: 1GFC, 2GFC, 4GFC, 8GFC, 16GFC	262
	9.3.5	Fibre over Ethernet (FCoE)	262
	9.3.6	InfiniBand (IBA)	264

9.3.7 Serial Attached SCSI (SAS) 265
9.3.8 Serial ATA (SATA) 267
9.3.9 TCP/IP 268
9.4 Abstracting Distance for Virtual Data Centers 269
9.4.1 Metropolitan and Wide Area Networks 269
9.4.2 Wide Area File Service (WAFS) and
Wide Area Application Service (WAAS) 274
9.5 Virtual I/O and I/O Virtualization 275
9.5.1 N_Port_ID Virtualization 277
9.5.2 Blade Center and Server Virtual
Convexity Features 279
9.5.3 Converged Networks 280
9.5.4 PCI-SIG IOV 282
9.5.5 Convergence Enhanced Ethernet
and FCoE 284
9.5.6 InfiniBand IOV 286
9.6 Virtualization and Management Tool Topics 287
9.6.1 Networking Options for Virtual
Environments 288
9.6.2 Oversubscription: Not Just for Networks 289
9.6.3 Security 290
9.6.4 Cabling and Cable Management 291
9.7 Summary 293

PART IV: Applying What You Have Learned 295

Chapter 10 Putting Together a Green and Virtual Data Center 297
10.1 Implementing a Green and Virtual Data Center 297
10.2 PCFE and Green Areas of Opportunity 300
10.2.1 Obtain and Leverage Incentives
and Rebates 301
10.2.2 Best Practices and IRM 301
10.2.3 Implement Metrics, Measurements 308
10.2.4 Mask-or-Move Issues 309
10.2.5 Consolidation 311
10.2.6 Reduced Data Footprint 312
10.2.7 Tiered Servers, Storage, and I/O
Network Access 316
10.2.8 Energy Avoidance—Tactical 318
10.2.9 Energy Efficiency—Strategic 319
10.2.10 Facilities Review and Enhancements 320

10.2.11 Environmental Health and Safety;
E-Waste; Recycle, Reuse, Reduce   320
10.3 Summary   321

Chapter 11   Wrap-up and Closing Comments   **323**

11.1 Where We Have Been   323
11.2 Where We Are Going—Emerging Technologies
and Trends   324
11.3 How We Can Get There—Best Practices and Tips   328
11.4 Chapter and Book Summary   329

Appendix A   Where to Learn More   **333**

Appendix B   Checklists and Tips   **337**

B.1 Facilities, Power, Cooling, Floor Space and
Environmental Health and Safety   337
B.2 Variable Energy Use for Servers   337
B.3 Variable Energy Use for Storage   338
B.4 Data Footprint Impact Reduction   339
B.5 Security and Data Protection   340
B.6 How Will Virtualization Fit into Your Existing
Environment?   341
B.7 Desktop, Remote Office/Branch Office (ROBO),
Small/Medium Business (SMB), and Small
Office/Home Office (SOHO) Users   341
B.8 Questions to Ask Vendors or Solution Providers 342
B.9 General Checklist and Tip Items   343

Glossary   **345**

Index   **367**

Preface

To say that "green" is a popular trend is an understatement. Green messaging in general and in the information technology (IT) industry in particular tends to center around carbon footprint and emissions reduction or cost savings. Green is also being seen or talked about as being dead or falling off in importance and relevance. While green hype and "green washing" may be falling out of favor or appearing on an endangered species list, addressing core IT data center issues that affect how resources are used to delivery information services in an energy-efficient, environmentally and economically friendly manner to boost efficiency and productivity is here to stay. There are many different aspects to actually being green as opposed to being perceived as being green. If, however, you listen closely, you might also hear mention of other topics and issues, including buzzwords such as RoHS, WEEE, LEED, J-MOSS, energy avoidance, and energy efficiency.

Common questions about green and IT include:

- Is green a consumer or enterprise public relations, science, or a political topic?
- Why are so few organizations going green with their business and IT organizations?
- Is green only an issue for large organizations that consume large amounts of energy?
- Is being green simply about reducing emissions or a new marketing tool?
- Is green only about cost savings?

- Is being green an inconvenient truth for IT organizations or vendors?

- Is green only about reducing energy emissions and carbon footprint?

- How does recycling and removal of hazardous substances fit into green themes?

- Is virtualization applicable only for server, storage, or network consolidation?

- Can existing IT environments transform in place to green and virtualized?

- What are the economic benefits of going green? What about the risks of doing nothing?

IT data centers around the world are faced with various power, cooling, floor space, and associated environmental health and safety issues while working to support growth without disrupting quality of service. This book sets aside the science of green and the political aspects of what is or is not considered green. Instead, this book looks at the various issues and opportunities for business and IT organizations that want to sustain growth in an economical and environmentally friendly manner.

IT infrastructure resources configured and deployed in a highly virtualized manner can be combined with other techniques and technologies to achieve simplified and cost-effective delivery of IT services to meet application service levels in an environmentally friendly and economical manner. If this resonates with you or you want to learn more, then this book will provide real-world perspectives and insight to addressing server, storage, networking, and facilities topics in a next-generation virtual data center that relies on an underlying physical infrastructure.

When I ask IT personal in the United States about green initiatives in their organizations, the response tends to indicate that about 5–10% of IT organizations have a green initiative, need, or mandate. However, the main response I find, in the 75–85% range, is centered around limits, constraints, or reliable availability of electrical power, or limits on cooling capacities and floor space.

Outside the United Statess, it is more common to hear about green initiatives in IT organizations, particularly when talking to people in Europe and parts of Asia where government mandates are either in place or being put into place. In the United States, and North America in general, such mandates are fragmented and confusing, and established on a state-by-state or regional basis.

At the heart of the growing "green gap" is messaging by the industry, or lack thereof, and the need for more awareness of IT organization core issues in different parts of the world or even, different parts of the United States. For example, the U.S. Environmental Protection Agency (EPA) is concerned about the environment, but it is also concerned about IT data centers having reliable flow and use of electricity. Hence, while Energy Star programs are about energy conservation and improved efficiency, they are also a means to help demand-side management. That is, the goal is to help ensure that demand for power will not out pace supply (generation and transmission), given the dense power requirements of IT data centers or information factories.

Several technologies exist now, and others are emerging, that can be used in complementary ways to enable a green and efficient virtual data center to support and sustain business growth. Some of the topics, technologies, and techniques that are covered in this book include energy and cost footprint reduction; cloud-based storage and computing; managed services; intelligent power management and adaptive power management; blade centers and blade servers; server, storage, and networking virtualization; data footprint reduction including archiving, compression, and data de-duplication; tiered servers; storage, network, and data centers; energy and environmentally friendly infrastructure resource management; and energy avoidance and energy efficiency; among others.

As the industry adjusts its messaging to the core needs of IT data centers in different geographies to address real issues of power, cooling, and floor space beyond basic cost reduction or power avoidance, there will be more success on the part of both vendors and IT organizations. Beyond being green, by enabling IT centers to do more with what they have (or even less), growth, performance, and service levels will be boosted, enabling improved efficiency and productivity while addressing both economic and environmental concerns.

Who Should Read This Book

This book cuts across various IT data technology domains as a single source to discuss the interdependencies that need to be supported to enable a virtualized, next-generation, energy-efficient, economical, and environmentally friendly data center. This book has been written with several audiences in mind, including IT purchasing, facilities, server, storage, networking, database, and applications analysts, administrators, and architects. Other audiences for this book include manufacturers and solution partners, sales, marketing, support, and engineering organizations, as well as public relations, investment communities, and media professionals associated with IT technologies and services.

How This Book Is Organized

This book is organized in four parts. Part I looks at IT data center economic and environmental issues and the many facets of being green, along with what comprises an energy-efficient and green data center. Part II looks at what defines a next-generation virtual data center, including infrastructure resource management and measurements to manage resources efficiently for improved productivity. Part III looks at the various technologies for enabling a green and virtual data center, including facilities, servers, storage, and networking, as well as associated infrastructure resource management. Finally, Part IV ties everything together, applying what you have learned. There are also various tips, sources of additional information, and checklists in the appendixes, along with a glossary.

Greg Schulz

Founder, the StorageIO Group

www.storageio.com

www.thegreenandvirtualdatacenter.com

Acknowledgments

Writing a book is more than putting pen to paper, or, in this case, typing on a computer. It takes many behind-the-scenes activities and people to bring a concept to reality. The author would like to thank all of those directly and indirectly involved with this project, including Andrew Fanara and Arthur (AJ) Howard of the U.S. Environmental Protection Agency, Dale Barth, Tom Becchetti, Eric Beastrom, Greg Brunton, Randy Cole, Jim Dyer, Jim Ellingson, Patrice Geyen, Scott Hale, David Hill, Rich Lillis, Keith Norbie, Bruce Ravid, Brad Severson, Marc Staimer, Louise Stich, George Terwey, and the Nelsons and Schoellers.

Thanks and appreciation to all of the vendors and channel professionals, service providers, members of the media, bloggers, writers, fellow analysts and consultants, industry trade groups, and of course all the IT professionals around the world whom I have been fortunate to meet and talk with while putting this book together.

Thanks to John Wyzalek, Jessica Vakili, and everyone else at CRC/Taylor Francis/Auerbach, and to Theron Shreve at Derrryfield Publishing Services and his crew, including Lynne Lackenbach and Tim Donar. Special thanks to Damaris Larson, who continues to push and challenge me in support of this and countless other projects.

Very special thanks to my wife Karen for having the patience and support not to mention her keeping "Big Babe" and "Little Leo" entertained during this project, as well as her family, as I wrote several chapters during our summer 2008 tour of the fjords, countryside, and cities of Norway, during which my father-in-law, Dr. Verlyn Anderson, was awarded the Norway's St. Olav Medal of Honor.

PART I: Green IT and the Green Gap-Real or Virtual?

PART II: Next-Generation Virtual Data centers

PART III: Technologies for Enabling Green and Virtual Data Centers

PART IV: Applying What You Have Learned

Chapter 1

IT Data Center Economic and Ecological Sustainment

Separating green washing and green issues in IT is the green gap.

In this chapter you will learn:

- The many faces of green information technology (IT) in data centers
- How to close the green gap and enable solutions to address IT issues
- IT data centers dependencies on electrical power
- Myths and realities pertaining to power, cooling, floor space, and environmental health and safety (PCFE)
- What green washing is and how to develop a PCFE strategy to address it
- Differences between avoiding energy use and being energy efficient
- Various techniques, technologies, and approaches to address PCFE issues

1.1   The Many Faces of Green—Environmental and Economic

In order to support business growth and ensure economic sustainability, organizations of all sizes need to establish a strategy to address the impact of their **power, cooling, floor space, and environmental health and safety (PCFE)** needs. PCFE addresses the many different facets of being green for **information technology (IT)** data centers, as shown in Figure 1.1, including emissions as a result of electricity energy production; cooling of IT equipment; efficient use of floor space; power for uninterruptible power supplies (UPS), backup, electricity **generation and transmission (G&T),** cost of power, and supply/demand; and disposal (removal of hazardous substances, waste electrical and electronic equipment, adherence to the LEED

3

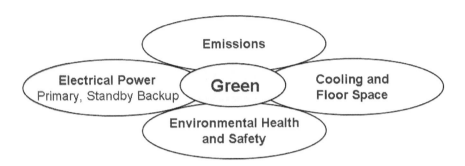

Figure 1.1   Green in IT Data Centers

[Leadership in Energy and Environmental Design] program). **Environmental health and safety (EHS)** topics include elimination of hazardous substances, traditional and e-waste disposal, and recycling. In addition to supporting growth, the business benefits include the abilities to leverage new and enhanced information services, enable business agility, and improve on cost effectiveness to remain competitive while reducing effects on the environment.

As a society, we have a growing reliance on storing more data and using more information-related services, all of which must be available when and where needed. Data and related information service are enabled via IT services including applications, facilities, networks, servers, and storage resources referred to collectively as an IT data center. As a result of this increasing reliance on information, both for home and personal use as well as business or professional use, more data is being generated, stored, processed, and retained for longer periods of time.

Energy costs are rising, floor space is at a premium in some organizations or will soon be exhausted in others. Cooling and electrical power distribution capabilities are strained or at their limits. Existing and emerging regulations for EHS as well as for emissions and energy efficiency are appearing. All of these factors continue to stress an aging infrastructure, which affects business growth and economic sustainment, resiliency, availability, and operations cost complexity. Meanwhile, businesses of all sizes have a growing dependence on availability and timely access to IT resources and data.

Telephone services, 911 emergency dispatches, hospitals and other critical services, including your favorite coffee house and the electrical power utilities themselves, rely on information services and the reliable supply of electrical power to sustain their existence. Not only is more data being generated, more copies of data are being made and distributed for both active use by information consumers and for data protection purposes, all resulting in an expanding data footprint. For example, when a popular document, presentation, photo, audio, or video file is posted to a website and subsequently downloaded, the result is multiple copies of the document being stored on different computer systems. Another common example of an expanding data footprint is the many copies of data that are made for business continuance, disaster recovery, or compliance and archiving purposes to protect and preserve data.

1.2   The Growing Green Gap: Misdirected Messaging, Opportunities for Action

The combination of growing demand for electricity by data centers, density of power usage per square foot, rising energy costs, strained electrical G&T infrastructure, and environmental awareness prompted the passage of U.S. Public Law 109-431 in 2006. Public Law 109-431 instructed the U.S. Environmental Protection Agency (EPA), part of the Department of Energy, to report to Congress on the state of IT data centers' energy usage in the United States.

In the August 2007 EPA report to Congress, findings included that IT data centers (termed information factories) in 2006 consumed about 61 billion kilowatt-hours (kWh), or 61 billion times 1,000 watt-hours of electricity, at an approximate cost of about $4.5 billion. It was also reported that IT data centers, on average, consume 15 to 25 times (or more) energy per square foot compared to a typical office building. Without changes in electricity consumption and improved efficiency, the EPA estimates that IT data centers' power consumption will exceed 100 billion kWh by 2011, further stressing an already strained electrical G&T infrastructure and increasing already high energy prices.

There is a growing "green gap" or disconnect between environmentally aware, focused messaging and core IT data center issues. For example, when I ask IT professionals whether they have or are under direction to implement green IT initiatives, the number averages below 20%. However, when I ask the same audiences who has or sees power, cooling, floor space, or EHS-related issues, the average is 80–90%. I have found some variances in

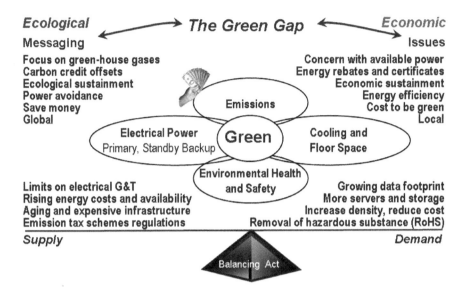

Figure 1.2   The Green Gap, Together with Supply and Demand Drivers

geography of IT people across the United States, as well as around the world, which changes the above numbers somewhat; however, the ratios are consistent. That is, most IT professionals relate being "green" with other items as opposed to making the link to PCFE topics. Not surprisingly, when I talk with vendors and others in the industry and ask similar questions, there is usually an inverse response, further indicating a green gap or messaging disconnects.

The growing green gap is, in its simplest terms, one of language and messaging. Although there is a common denominator or linkage point of green among data center issues, there is often a disconnect between needs centered on PCFE footprint and related costs as well as the need to sustain business growth. Building on Figure 1.1, Figure 1.2 shows common environmental messaging on the left and common IT issues shown on the right. Also in Figure 1.2, on the lower left, are supply constraints, with some demand drivers shown in the lower right. Specific issues or combinations of issues vary by organization size, location, reliance on and complexity of IT applications, and servers, among other factors.

IT organizations in general have not been placing a high priority on being perceived as green, focusing instead on seemingly nongreen PCFE issues. Part of the green gap is that many IT organizations are addressing (or need to address) PCFE issues without making the connection that, in fact;

they are adopting green practices, directly or indirectly. Consequently, industry messaging is not effectively communicating the availability of green solutions to help IT organizations address their issues. By addressing IT issues today that include power, cooling, and floor space along with asset disposal and recycling, the by-products are economic and ecologically positive. Likewise, the shift in thinking from power avoidance to more efficient use of energy helps from both economic and ecological standpoints.

There is some parallel between the oil crisis of the 1970s and the current buzz around green IT and green storage along with power, cooling, and floor space issues. During the 1970s oil crisis, there was huge pressure to conserve and avoid energy consumption, and we are seeing similar messaging around power avoidance for IT, including consolidation and powering down of servers and storage systems.

Following the initial push for energy conservation in the 1970s was the introduction of more energy-efficient vehicles. Today, with IT resources, there is a focus on more energy-efficient servers and storage, for both active and inactive use or applications, which also incorporate intelligent power management or adaptive power management along with servers and storage that can do more work per watt of energy. The subsequent developments involve adopting best practices including better data and storage management, archiving, compression, de-duplication, and other forms of data footprint reduction, along with other techniques to do more with available resources to sustain growth.

1.3   IT Data Center "Green" Myths and Realties

Is "green IT" a convenient or inconvenient truth or a legend? When it comes to green and virtual environments, there are plenty of myths and realities, some of which vary depending on market or industry focus, price band, and other factors. For example, there are lines of thinking that only ultralarge data centers are subject to PCFE-related issues, or that all data centers need to be built along the Columbia River basin in Washington state, or that virtualization eliminates vendor lock-in, or that hardware is more expensive to power and cool than it is to buy. The following are some myths and realities as of today, some of which may be subject to change from reality to myth or from myth to reality as time progresses.

- *Myth:* Green and PCFE issues are applicable only to large environments.

  - *Reality:* I commonly hear that green IT applies only to the largest of companies. The reality is that PCFE issues or green topics are relevant to environments of all sizes, from the largest of enterprises to the small/medium business, to the remote office branch office, to the small office/home office or "virtual office, all the way to the digital home and consumer.

- *Myth:* All computer storage is the same, and powering disks off solves PCFE issues.

  - *Reality:* There are many different types of computer storage, with various performance, capacity, power consumption, and cost attributes. Although some storage can be powered off, other storage that is needed for online access does not lend itself to being powered off and on. For storage that needs to be always online and accessible, energy efficiency is achieved by doing more with less, that is, boosting performance and storing more data in a smaller footprint using less power.

- *Myth:* Servers are the main consumer of electrical power in IT data centers.

  - *Reality:* In the typical IT data center, on average, 50% of electrical power is consumed by cooling, with the balance used for servers, storage, networking, and other aspects. However, in many environments, particularly processing or computation-intensive environments, servers in total (including power for cooling and to power the equipment) can be a major power draw.

- *Myth:* IT data centers produce 2% of all global carbon dioxide (CO_2) emissions.

  - *Reality:* This is perhaps true, given some creative accounting and marketing math. The reality is that in the United States, for example, IT data centers consume around 2–3% of electrical power (depending on when you read this), and less than 80% of all U.S. CO_2 emissions are from electrical power generation, so the math does not quite add up. However, if no action is taken to improve IT data center energy efficiency, continued demand growth will shift IT power-related emissions from myth to reality.

- *Myth:* Server consolidation with virtualization is a silver bullet to address PCFE issues.
  - *Reality:* Server virtualization for consolidation is only part of an overall solution that should be combined with other techniques, including lower power, faster and more energy-efficient servers, and improved data and storage management techniques.

- *Myth:* Hardware costs more to power than to purchase.
  - *Reality:* Currently, for some low-cost servers, standalone disk storage, or entry-level networking switches and desktops, this may be true, particularly where energy costs are excessively high and the devices are kept and used continually for three to five years. A general rule of thumb is that the actual cost of most IT hardware will be a fraction of the price of associated management and software tool costs plus facilities and cooling costs.

Regarding this last myth, for the more commonly deployed external storage systems across all price bands and categories, generally speaking, except for extremely inefficient and hot-running legacy equipment, the reality is that it is still cheaper to power the equipment than to buy it. Having said that, there are some qualifiers that should also be used as key indicators to keep the equation balanced. These qualifiers include the acquisition cost; the cost, if any, for new, expanded, or remodeled habitats or space to house the equipment; the price of energy in a given region, including surcharges, as well as cooling, length of time, and continuous time the device will be used.

For larger businesses, IT equipment in general still costs more to purchase than to power, particularly with newer, more energy-efficient devices. However, given rising energy prices, or the need to build new facilities, this could change moving forward, particularly if a move toward energy efficiency is not undertaken.

There are many variables when purchasing hardware, including acquisition cost, the energy efficiency of the device, power and cooling costs for a given location and habitat, and facilities costs. For example, if a new storage solution is purchased for $100,000, yet new habitat or facilities must be built for three to five times the cost of the equipment, those costs must be figured into the purchase cost. Likewise, if the price of a storage solution decreases dramatically, but the device consumes a lot of electrical power and needs a large cooling capacity while operating in a region with expensive electricity costs, that, too, will change the equation and the potential reality of the myth.

1.4   PCFE Trends, Issues, Drivers, and Related Factors

Core issues and drivers of green IT include:

- Public consumer and shareholder or investor environmental pressures
- Recycling, reducing, and reusing of resources, including safe technology disposition
- Increasing demand and reliance on information factories and available IT resources
- Limited supply and distribution capabilities for reliable energy
- Pressures from investors and customers to reduce costs and boost productivity
- Constraints on power distribution, cooling capabilities, and standby power sources
- Rising energy and distribution costs coupled with availability of energy fuel sources
- Growing global environmental and green awareness and messaging
- Enforcement of existing clean air acts along with reduction of energy-related emissions:
  - Greenhouse gasses, including CO_2 and nitrogen dioxide (NO_2)
  - Methane, water vapor, acid rain, and other hazardous by-products
- Emerging and existing EHS regulations or legislation standards and guidelines:
  - Enhancements to existing clean air act legislation

- The Kyoto Protocol for climate stability
- Emission tax schemes in Europe and other regions
- Removal of hazardous substances
- Waste electrical and electronic equipment
- U.S. Energy Star, European Union (EU), United Kingdom (UK) programs

The U.S. national average CO_2 emission is 1.34 lb/kWh of electrical power. Granted, this number will vary depending on the region of the country and the source of fuel for the power-generating station or power plant. Coal continues to be a dominant fuel source for electrical power generation both in the United States and abroad, with other fuel sources, including oil, gas, natural gas, liquefied propane gas (LPG or propane), nuclear, hydro, thermo or steam, wind and solar. Within a category of fuel for example, coal there are different emissions per ton of fuel burned. Eastern U.S. coal is higher in CO_2 emissions per kilowatt-hour than western U.S. lignite coal. However, eastern coal has more British thermal units (Btu) of energy per ton of coal, enabling less coal to be burned in smaller physical power plants.

If you have ever noticed that coal power plants in the United States seem to be smaller in the eastern states than in the midwestern and western states, it's not an optical illusion. Because eastern coal burns hotter, producing more Btu, smaller boilers and stockpiles of coal are needed, making for smaller power plant footprints. On the other hand, as you move into the midwestern and western states of the United States, coal power plants are physically larger, because more coal is needed to generate 1 kWh, resulting in bigger boilers and vent stacks along with larger coal stockpiles.

On average, a gallon of gasoline produces about 20 lb of CO_2, depending on usage and efficiency of the engine as well as the nature of the fuel in terms of octane or amount of Btu. Aviation fuel and diesel fuel differ from gasoline, as do natural gas or various types of coal commonly used in the generation of electricity. For example, natural gas is less expensive than LPG but also provides fewer Btu per gallon or pound of fuel. This means that more natural gas is needed as a fuel to generate a given amount of power.

Recently, while researching small, 10- to 12-kWh standby generators for my office, I learned about some of the differences between propane and natural gas. What I found was that with natural gas as fuel, a given generator produced about 10.5 kWh, whereas the same unit attached to a LPG or propane fuel source produced 12 kWh. The trade-off was that to get as much power as possible out of the generator, the higher-cost LPG was the better choice. To use lower-cost fuel but get less power out of the device, the choice would be natural gas. If more power was needed, than a larger generator could be deployed to use natural gas, with the trade-off of requiring a larger physical footprint.

Oil and gas are not used as much as fuel sources for electrical power generation in the United States as in other countries such as the United Kingdom. Gasoline, diesel, and other petroleum-based fuels are used for some power plants in the United States, including standby or peaking plants. In the electrical power G&T industry as in IT, where different tiers of servers and storage are used for different applications there are different tiers of power plants using different fuels with various costs. Peaking and standby plants are brought online when there is heavy demand for electrical power, during disruptions when a lower-cost or more environmentally friendly plant goes offline for planned maintenance, or in the event of a "trip" or unplanned outage.

CO_2 is commonly discussed with respect to green and associated emissions. Carbon makes up only a fraction of CO_2. To be specific, only about 27% of a pound of CO_2 is carbon; the balance is not. Consequently, carbon emissions taxes, as opposed to CO_2 tax schemes, need to account for the amount of carbon per ton of CO_2 being put into the atmosphere. In some parts of the world, including the EU and the UK, **emission tax schemes (ETS)** are either already in place, or are in initial pilot phases, to provide incentives to improve energy efficiency and use. Meanwhile, in the United States there are voluntary programs for buying carbon offset credits as well as initiatives such as the Carbon Disclosure Project. The Carbon Disclosure Project (www.cdproject.net) is a not-for-profit organization that facilitates the flow of information about emissions to allow managers and investors to make informed decisions from both economic and environmental perspectives. Another voluntary program is the EPA Climate Leaders initiative, under which organizations commit to reduce their greenhouse gas emissions to a given level or within a specific period of time.

These voluntary programs can be used to offset carbon emissions. For example, through organizations such as Terrapass (www.terrapass.com), companies or individuals can buy carbon credits to offset emissions from home or business energy use. Credits can also be bought to offset emissions associated with travel or automobile use. Money from the sale of carbon offset credits is applied to environmental clean-up, planting of trees, and other ways of reducing climate change. Carbon credits are also traded like traditional securities and commodities via groups such as the Chicago Climate Exchange (www.chicagoclimateexchange.com). One way of thinking about carbon credits is to consider them the way you would a parking or speeding ticket: What would it take (stiffer fines, jail time) to make you change your driving habits?

A controversial use of carbon credits is associated with green washing," the practice of using carbon credits to offset or even hide energy use or lack of energy efficiency. For example, if many individuals fly on their own or charter jet aircraft as opposed to flying on commercial aircraft, on which emissions are amortized across more passengers, carbon offsets are a convenient way of being perceived as carbon neutral. For a relatively older plane such the Boeing 757, a late 1970s-era twin-engine design, which can carry about 150 passengers and crew on a 1,500-mile flight, the average fuel used per passenger for a flight of just over three hours is in the range of 65 mpg per passenger. For a larger and newer more fuel-efficient aircraft, which can carry more passengers over longer distances, the average miles per gallon per passenger may be closer to 75 mpg if the plane is relatively full. On the other hand, if the same airplane is half-empty, the miles per gallon per passenger will be lower.

If you are concerned about helping the environment, there are options including planting trees, changing your energy usage habits, and implementing more energy-efficient technology. A rough guideline is that about one forested acre of trees is required to offset 3 tons of CO_2 emissions from electrical power generation. Of course, the type of tree, how densely the acre of trees is planted, and the type of fuel being used to produce electrical power will alter the equation. Energy costs vary by region and state as well as from residential to business, based on power consumption levels. Some businesses may see a base rate of, for example, 12 cents/kWh used, which jumps to 20 cents/kWh after some number of kWh used each month.

Energy costs also need to reflect cooling costs, which typically account for about half of all power consumed by IT data centers. A typical gallon of gasoline (depending on octane level) will, on average, generate about 20 lb of CO_2. Approximately (this number is constantly changing) 78% of CO_2 emissions in the United States are tied to electrical power generation. Carbon offset credits can sell for in the range of \$250/lb up to 10 tons of CO_2 and \$200/lb for larger quantities up to 35 tons of CO_2.

The importance of these numbers and metrics is to focus on the larger impact of a piece of IT equipment that includes in its cost not only energy consumption but also other hosting or site environmental costs. Energy costs and CO_2 emissions vary by geography and region, as does the type of electrical power used (coal, natural gas, nuclear, wind, thermal, solar, etc.) and other factors that should be kept in perspective as part of the big picture.

Some internal and external motivators driving awareness and discussion of green IT and the need to address PCFE issues include:

- Family and philanthropic motive versus economic business realties
- Debates between science and emotions, and the impact of politics and business realities
- Cost savings or avoidance, or when to spend money to save larger amounts of money
- Evolving awareness of green global supply chains and ecosystems

Internal and external stakeholders or concerned entities include:

- Customers and employees concerned about EHS or energy savings
- Public concern about climate change, recycling, and removal of hazardous substances
- Investors and financial communities looking to avoid risk and liabilities
- Governmental and regulatory bodies enforcing existing or emerging legislation
- General public relations to protect and preserve a businesss image and reputation

Given the many different faces of green and PCFE issues, regulations and legislation vary by jurisdiction and areas of focus. For example, under the Kyoto Protocol on climate change control, not all nations have indicated how they will specifically reduce emissions. Some countries and regions, including the UK, the EU, and Australia, either have deployed or are in the process of launching ETS programs. In the United States there are voluntary programs for buying carbon emissions offsets along with energy-efficiency incentives programs. There is, however, as of this writing, no formal country-wide ETS.

1.5   Closing the Green Gap for IT Data Centers

In order to support the expanding data footprint and reliance on timely information services, more servers, storage, networks, and facilities are required to host, process, and protect data and information assets. IT data centers, also known as information factories, house, process, and protect the information we depend on in our daily lives.

A common challenge for enabling continued access to information faced by IT data centers of all sizes, regardless of type, business, or location, is cost containment. Over the past decades, technology improvements have seen computers reduced to a fraction of the physical size, weight, cost, and power consumption formerly needed, while providing an exponential increase in processing capability. Data storage technologies, including magnetic disk drives, have also seen a similar decline in physical size, weight, cost, and power consumption, while improving on available storage capacity, reliability, and, in some cases, performance. Likewise, networks have become faster while requiring a smaller footprint, both physically and in terms of power required to operate the equipment. Even with technology improvements, however, the demand and subsequent growth rate for information-related services are outpacing the supply of reliable and available PCFE footprints.

For example, wake up in the morning and turn on the news, on the radio, TV, or Internet, and you are relying on the availability of information resources. The newspaper that you may read, your email, Instant Messages (IM) or text messages, voice mail, phone calls regardless of whether on a traditional landline phone, cell phone, or Internet-based phone service rely on information services. In the workplace, reliance on information services becomes more apparent, ranging from office to factory environments; travel depends on reservations, scheduling, weights and balance, dispatch and control systems. These are just a few examples of our dependence on available information. A few more examples include reliance on information services when you go shopping, whether online or in person at a large megastore store, or to the corner market; regardless of size, there are credit card processing, barcode labels or radio frequency ID (RFID) tags to read at checkout and for inventory.

Throughout this book, additional examples will be presented. However, for now, one last example of our growing reliance and use of information services is how much personal data exists in your home. That is, add up the amount of data, including digital photos, iTunes or MP3 audio files such as CDs and DVDs, either "ripped" onto a computer disk drive or in their native format, other documents on your personal digital assistants (PDAs), cell phones, and TiVo or digital video recorders (DVRs). For those who have a computer at home along with a cell phone and a digital camera, it is very likely that there exists at least 100 gigabytes (GB) of personal information that is, 100,000,000,000 characters of information. For many households, that number is closer to if not already exceeding 1,000 GB or 1 terabyte of data or 1,000,000,000,000 characters of information.

Even with the advent of denser IT equipment that does more in a given footprint, the prolific growth rate and demand for existing along with new rich media applications and data outpace available PCFE footprints. The result is a growing imbalance between supply of affordable and reliable power distribution where needed and the increasing demand and reliance on information services. Put another way, organizations that rely on information services need to sustain growth, which means more data being generated, processed, stored, and protected, which has a corresponding PCFE footprint impact.

IT services consumers are increasingly looking for solutions and products that are delivered via companies with green supply chains and green ecosystems. Green supply chains extend beyond product logos with green backgrounds or pictures of green leaves on the packaging to simply make you feel good about going green.

Regardless of stance or perception on green issues, the reality is that for business and IT sustainability, a focus on ecological and, in particular, the corresponding economic aspects cannot be ignored. There are business benefits to aligning the most energy-efficient and low-power IT solutions combined with best practices to meet different data and application requirements in an economic and ecologically friendly manner.

For example, by adopting a strategy to address PCFE issues, benefits include:

- Business and economic sustainment minimizing risk due to PCFE compliance issues
- Avoidance of economic loss due to lack of timely and cost-effective growth
- Reduction in energy-related expenses or downtime due to loss of power
- Compliance with existing or pending EHS legislation or compliance requirements
- Realization of positive press and public relations perception by consumers and investors
- Maximization of investment in facilities, floor space, cooling, and standby power
- Increasing business productivity and value with enhanced information systems
- The ability to adapt to changing economic and business conditions in a timely manner
- Leveraging of information systems as a core asset as opposed to a business cost center

As consumers in general are becoming aware of the many faces and issues associated with being green, there is also a growing skepticism about products, services, and companies that wrap themselves in green marketing. This has lead on a broad basis well beyond just the IT industry to the practice of "green washing, which involves "painting" a message "green" to appeal to the growing global green and environmental awareness phenomena. The challenge with "green washing" is that viable messages and important stories too often get lost in the noise, and, subsequently, real and core issues or solutions are dismissed as more "green noise. The "green washing" issue has become such a problem that traditional green-focused organizations such as Greenpeace have launched campaigns with websites (see, e.g., www.stopgreenwash.org) against the practice of "green washing" with slogans such as "Clean up your act, NOT your image."

Green initiatives need to be seen in a different light, as business enablers as opposed to ecological cost centers. For example, many local utilities and state energy or environmentally concerned organizations are providing funding, grants, loans, or other incentives to improve energy efficiency. Some of these programs can help offset the costs of doing business and going green. Instead of being seen as the cost to go green, by addressing efficiency, the by-products are economic as well as ecological. Put a different way, a company can spend carbon credits to offset its environmental impact, similar to paying a fine for noncompliance, or it can achieve efficiency and obtain incentives. There are many solutions and approaches to address these different issues, which will be looked at in the coming chapters.

Green and other energy, EHS, or PCFE or environmental power and cooling issues are a global concern for organizations of all sizes as well as for individuals. Issues and areas of focus and requirements vary by country and region. Some countries have more stringent regulations, such as specific guidelines for reducing energy consumption and CO_2 emissions or paying carbon offset credits, whereas in other countries the focus is on reducing costs or matching demand with energy availability.

Simply put, addressing PCFE issues or being green is an approach and practice for acquiring, managing, and utilizing IT resources to deliver application and data services in an economic and ecologically friendly manner for business sustainment. In going or being green, by improving IT infrastructure and resource efficiency, doing more with less, and maximizing existing PCFE and energy to become more ecological friendly, you also enable a business to grow, diversify, and expand its use of IT, all of which have economic benefits.

For example, you can spend money to become green by buying carbon offset credits while you continue to operate IT resources including servers, storage, and networks in an inefficient manner, or you can improve your efficiency by consolidating, boosting performance to do more work per unit of energy, reducing your PCFE impact and associated costs, and thus creating an economic benefit that also benefits the environment.

A challenge in identifying and addressing how effective solutions and approaches are toward being green is the lack of consistent or standard measurements, metrics, and reporting. For example, vendors may list energy used by hardware devices in many different ways, including watts, Btu, or amps for either the maximum circuit load, idle, active, or some other state such as a maximum configuration and workload. Some are measured; some are estimated or derived from a combination of measurements and component estimates.

Another factor tied to measuring energy use is how to gauge the effectiveness of the energy being used. For example, energy used for a tape drive and the tape is different than the energy used for a high performance online storage system. Even when supposedly turned off or power down, many IT devices and consumer electronics still draw some power. Granted, the amount of power consumed in a standby mode may be a fraction of what is used during active use, but power is still being consumed and heat generated. For example, a server can be put into a sleep or low-power mode using a fraction of the electrical power needed when performing work. Another example is consumer electronics such as digital TVs that, even when turned off, still consume a small amount of electrical power. On the basis of an individual household, the power used is small; on a large-scale basis, however, the power consumption of standby mode use should be factored into analysis along with any subsequent power-on spikes or surges.

These issues are not unique to IT. There are good and interesting similarities to the automobile and transportation sectors in terms of benchmarks, metrics, usage scenarios, and classifications of vehicles. Another interesting correlation is that the automobile industry had to go into a conservation mode in the 1970s during the oil embargo, when supply could not keep up with demand. This led to more energy-efficient, less polluting vehicles, leading to where we are today, with hybrid technologies and better metrics (real-time for some vehicles) and fuel options and driving habits.

Another corollary is the trend to over consolidate to boost utilization at the expense of effective service delivery. For example, in some IT environments, servers are being consolidated with a focus on boosting utilization to avoid energy use independent of the subsequent effect on performance or availability. The same can be said for the trend toward shifting away from underutilized large vehicles to smaller, energy-efficient, and so-called green hybrid vehicles. For both automobiles and IT, some consolidation is needed, as are more energy-efficient solutions, better driving and usage habits, and alternative energy and fuel sources. However, for both automobiles and IT, the picture needs to be kept in focus: Performance, availability, capacity, and energy consumption need to be balanced with particular usage needs. Put another way, align the applicable resource to the task at hand.

While businesses generally want to do what is good, including what is good for the environment (or at least put up a good story), the reality is that its hard cold economics, particularly in the absence of regulations, that dictate how business operate. This is where the green gap exists between going green to be green or to save money, as opposed to achieving and maintaining economic growth while benefiting the environment. In addressing business economics and operations to avoid bottlenecks and expenses while also helping the environment, alternatives that happen to be green are seen as more appealing and affordable.

1.5.1   Energy Consumption and Emissions: Green Spotlight Focus

The advent of cheaper volume computer power has brought with it a tremendous growth in data storage. Until recently, the energy efficiency of servers, networks, software, and storage had been of little concern to IT organizations. This is changing as the price of electricity is steadily increasing and demand is outpacing the supply of electricity G&T capabilities on both local and global bases. As application server utilization and energy efficiency for powering and cooling these IT resources improve, the focus will expand to include data storage and networking equipment.

A reliable supply of electricity is becoming more difficult to guarantee, because of finite G&T capacity and rising fuel costs. Added to this is increasing demand as a result of proliferating data footprints, more and denser servers, storage, and networks, along with limited floor space, backup power, and cooling capacity. As IT data centers address power, cooling, and floor space challenges with improved energy efficiency and effectiveness, three main benefits will be realized: helping the environment, reducing power and cooling costs, and enabling sustained application growth to support evolving business information needs.

Initially, power and cooling issues are being focused on in larger environments because of their size and scale of energy consumption; however, all environments should be sensitive to energy consumption moving forward. The following examples vary by location, applications, IT equipment footprint, power consumption, budgets, and other factors. For example, in Figure 1.3, the amount of energy required to cool and power IT equipment is shown along with the relative increase reflecting faster, more powerful, and denser equipment. In Figure 1.3, to the right of the graph is shown an area where PCFE resources are constrained, inhibiting growth or causing

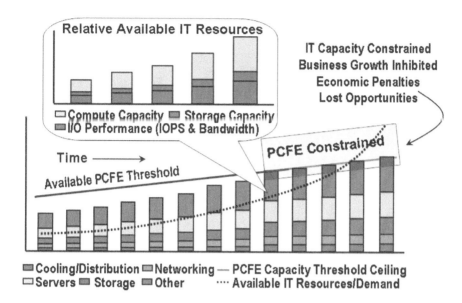

Figure 1.3   IT Resource Growth Constrained by PCFE Issues

increased costs to offset higher energy costs or to pay ETS-related fees, if applicable. Note in Figure 1.3 that, over time, IT equipment is becoming more efficient; however, the sheer density to support increasing demands for IT resources is putting a squeeze on available PCFE resource footprints.

In Figure 1.4, an example is shown with improvements in how energy is used, including deployment of more efficient servers, storage, and networking equipment capable of processing more information faster and housing larger amounts of data. The result of doing more work per watt of energy and storing more data in a given footprint per amount of energy used is that more PCFE resources are made available. The available PCFE footprint enables organizations either to reduce costs and their associated environmental footprint or to sustain business growth while enabling some PCFE resources to be used to transition from older, less efficient technology to newer, more energy-efficient technologies. Additional benefits can be achieved by combining newer technologies with improved data and storage management tools and best practices to further maximize PCFE resources.

1.5.2   EHS and Recycling: The Other Green Focus

Another important facet or face of green is EHS and recycling activities. Recycling and health and safety programs are hardly new, having been

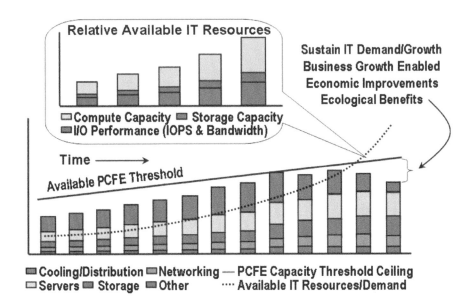

Figure 1.4   Sustaining Business and IT Growth by Maximizing PCFE Resources

around, in some cases, for a few decades. In fact, there are more existing regulations and legislation pertaining to EHS, recycling, disposition, and removal of hazardous substances on a global basis than there are to emissions and carbon footprints. For example, as of this writing, the U.S. Senate is beginning debate on climate change and emissions, while the EU and UK, along with other countries, have ETS programs in place.

The United States does have the Clean Air Act, dating back to the late 1960s and early 1970s, which dictates how certain pollutants are to be reduced, contained, and eliminated. Another example is various material handling standards and regulations, including for **removal of hazardous substances (RoHS)** such as bromine, chlorine, mercury, and lead from IT equipment, which vary with different country implementations. Other examples include politics and regulations around waste recycling as well as reduction and reuse of water or other natural resources in an environmentally friendly manner.

1.5.3   Establishing a Green PCFE Strategy

Do you have a PCFE or green strategy, and, if so, what drivers or issues does it address? If you do not yet have a PCFE strategy or are in the process of

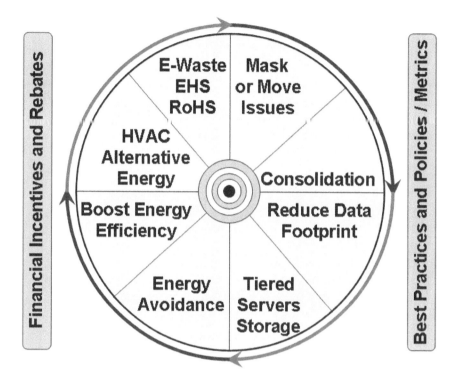

Figure 1.5   Techniques, Best Practices, and Technologies to Address PCFE Issues

creating one, some initial steps include identifying the many different issues. Different issues have varying impacts and scope, depending on specific plan points or requirements along with several alternatives for addressing on a near-term tactical and longer-term strategic basis. Several approaches are shown in Figure 1.5.

Figure 1.5 shows what I refer to as the PCFE wheel of opportunity. The various tenets can be used separately or in combination to address near-term tactical or long-term strategic goals. The basic premise of the PCFE wheel is to improve energy efficiency, by doing more with less, and/or boosting productivity and service levels to maximize IT operating and capital yields. For example, by maximizing energy efficiency, more work can be done with highly energy-efficient servers and storage that process more transactions, IOPS (input/output operations per second), or bandwidth per watt of energy.

Another possibility is consolidation. Underutilized servers and storage that lend themselves to consolidation can be aggregated to free servers for

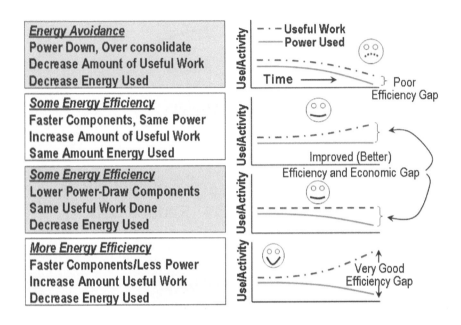

Figure 1.6   The Many Faces of Energy Efficiency

growth or to save energy. The end result is that, by achieving energy efficiency, IT costs may be lowered or, more likely, IT spending, including for electrical power, can be maximized to sustain business and economic growth in an environmentally friendly manner.

Consolidation has been a popular approach to address energy consumption; however, caution should be exercised, particularly with primary external storage. Over consolidation to drive capacity utilization can result in negative impacts on performance.

One point to remember is that in the typical IT data center, depending on configuration, application workload, and **service-level agreements (SLAs),** among other factors, only a small percentage of the overall server, storage, and networking resources will lend themselves to consolidation. Applications that need more processing or server performance capacity, storage, or IT resources beyond what a single device can provide typically are not candidates for consolidation. Another example is situations where different customers, clients, departments, or groups of users need to be isolated for security, financial, political, or other reasons. Servers or storage systems that can only be configured to use a certain amount of resources to meet performance response-time SLAs also are not candidates for consolidation.

Estimates range from as low as 5% to as high as 30% of servers that can be safely consolidated without sacrificing performance or negatively affecting service levels. Estimates on storage utilization vary even more, given the multiples ways storage can be configured and data stored to achieve performance or support operational functions. Not surprisingly, vendors of storage utilization improvement technology, including consolidation virtualization technology or storage resource management and capacity planning software, will talk about storage being only 15-30% allocated and used, on average. On the other hand, finding and talking to IT organizations that operate their open systems storage beyond 50-75% is not uncommon.

Figure 1.6 shows the many aspect of energy efficiency in general terms, with a major emphasis on maximizing available electrical power, improving energy efficiency, and doing more with less to contain electrical power-related emissions. In later chapters, various approaches, techniques, and technologies will be looked at from facilities, server, and storage and networking perspectives to address PCFE issues, including situations where consolidation is not an option.

Given the reliance on available electrical energy to both power and cool IT equipment and the emissions that are a by-product of generating elec-

Table 1.1   Energy Efficiency Scenarios

Energy Scenario	Description and Usage
Energy avoidance	Avoid or decrease the amount of work or activity to be done to reduce energy usage. On the surface, it is appealing to simply turn things off and avoid using energy. For some items, such as lights or video monitors or personal computers, this is a good practice. However, not all IT resources lend themselves to being turned off, as they need to remain powered on to perform work when and where needed.
Do the same work with less energy	Reduce energy usage, performing the same amount of work or storing the same amount of data per unit of energy used with no productivity improvement. This can be a stepping stone to doing more with less.

Table 1.1   Energy Efficiency Scenarios (continued)

Do more work with existing energy	Fit into an existing available energy footprint while doing more useful work or storing more data to improve efficiency. Although energy usage does not decline, energy efficiency is achieved by boosting the amount of work or activity performed for a given amount of energy used.
Do more work with less energy	Reduce energy consumption by boosting productivity and energy efficiency, doing more work, storing more information in a given footprint, using less energy and related cooling. An analogy is to improve miles per gallon for a distance traveled to boost energy efficiency for active work.

tricity, energy is a common area of green focus for many organizations. Improving energy usage or boosting energy efficiency can take on different meanings, as shown in Figure 1.6.

In Figure 1.6 are shown four basic approaches (in addition to doing nothing) to energy efficiency. One approach is to avoid energy usage, similar to following a rationing model, but this approach will affect the amount of work that can be accomplished. Another approach is to do more work using the same amount of energy, boosting energy efficiency, or the complement—do the same work using less energy. Both of these approaches, expanded on in Table 1.1, improve the energy efficiency gap. The energy efficiency gap is the difference between the amount of work accomplished or information stored in a given footprint and the energy consumed. In other words, the bigger the energy efficiency gap, the better, as seen in the fourth scenario, doing more work or storing more information in a smaller footprint using less energy.

Avoiding energy usage by turning off devices when they are not needed is an intuitive approach to reducing energy usage. For example, turning off or enabling intelligent power management for monitors or desktop servers should be as automatic as turning off overhead lights when they are not needed. On the other hand, turning off larger and mission-critical servers and associated active storage systems can have a negative impact on application availability and performance. Consequently, avoiding power usage is not typically a binary on/off solution for most data center environments. Instead, selectively powering down after analyzing application interdependences and associated business impacts or benefits should be pursued.

1.6   Summary

There are real things that can be done today that can be effective toward achieving a balance of performance, availability, capacity, and energy effectiveness to meet particular application and service needs. Sustaining for economic and ecological purposes can be achieved by balancing performance, availability, capacity, and energy to applicable application service-level and physical floor space constraints along with intelligent power management. Energy economics should be considered as much a strategic resource part of IT data centers as are servers, storage, networks, software, and personnel.

The IT industry is shifting from the first wave of awareness and green hype to the second wave of delivering and adopting more efficient and effective solutions. However, as parts of the industry shift toward closing the green gap, stragglers and late-comers will continue to message and play to the first wave themes, resulting in some disconnect for the foreseeable future. Meanwhile, a third wave, addressing future and emerging technologies, will continue to evolve, adding to the confusion of what can be done today as opposed to what might be done in the future.

The bottom line is that without electrical power, IT data centers come to a halt. Rising fuel prices, strained generating and transmission facilities for electrical power, and a growing awareness of environmental issues are forcing businesses to look at PCFE issues. IT data centers to support and sustain business growth, including storing and processing more data, need to leverage energy efficiency as a means of addressing PCFE issues. By adopting effective solutions, economic value can be achieved with positive ecological results while sustaining business growth.

Depending on the PCFE-related challenges being faced, there are several general approaches that can be adopted individually or in combination as part of a green initiative to improve how a business is operated and its goods or services delivered to customers.

General action items include:

- Learn about and comply with relevant environmental health and safety regulations.
- Participate in recycling programs, including safe disposal of e-waste.

- Reduce greenhouse gases, CO_2, NO_2, water vapor, and other emissions.

- In lieu of energy-efficient improvements, buy emissions offset credits for compliance.

- Comply with existing and emerging emissions tax schemes legislation and clean air acts.

- Increase awareness of energy and IT productivity and shift toward an energy-efficient model.

- Improve energy efficiency while leveraging renewable or green energy sources.

- Mask or move PCFE by buying emissions offsets, outsourcing, or relocating IT facilities.

- Identify differences between energy avoidance and efficiency to boost productivity.

- Archive inactive data; delete data no longer needed for compliance or other uses.

Chapter 2

Energy-Efficient and Ecologically Friendly Data Centers

Pay your speeding tickets and parking fines or spend your savings elsewhere.

In this chapter you will learn:

- How to identify issues that affect the availability of power, cooling, and floor space
- Why achieving energy efficiency is important to sustain growth and business productivity
- How electrical power is generated, transmitted, and used in typical data centers
- How electrical power is measured and charges determined

By understanding fundamentals and background information about electricity usage as well as options and alternatives including rebates or incentives, IT data centers can deploy strategies to become more energy-efficient without degrading service delivery. Reducing carbon footprint is a popular and trendy topic, but addressing energy efficiency—that is, doing more work with less energy—addresses both environmental and business economic issues. The importance of this chapter is that near-term economic as well as environmental gains can be realized by making more efficient use of energy. By reducing energy consumption or shifting to a more energy-efficient IT model, businesses can reduce their operating expenses and enable more useful work to be done per dollar spent while improving service delivery. This chapter looks at challenges with electrical power for IT data centers as well as background information to help formulate effective strategies to become energy-efficient.

2.1   Electric Power and Cooling Challenges

Asking the right questions can help you to close the "green gap" and address **power, cooling, floor space, and environmental (PCFE)** issues. That is, insight into how infrastructure resources are being used to meet delivery and service levels is critical. For example, instead of asking whether there is a green mandate or initiative, try asking the following questions:

- Does the data center have a power issue or anticipate one in the next 18–24 months?
- Does the IT data center have enough primary and backup power capacity?
- Is there enough cooling capacity and floor space to support near-term growth?
- How much power does the data center consume?
- How much of that power goes for cooling, lighting, and other facility overhead items?
- How much power is used by servers, storage, and networking components?
- Is power constrained by facility, local substation, or generating capability limits?
- What floor space constraints exist, and is there adequate cooling capabilities for growth?
- Can energy usage be aligned with the level of service delivered or amount of data stored?
- What hazardous substances and materials exist in the data center?

Closing the green gap is important in that core IT PCFE issues can be addressed with positive environmental and economic benefits. For example, building on the previous questions, common PCFE-related pain points for many IT data centers include:

- A growing awareness of green and environmentally friendly issues and topics

- The need to remove heat from IT equipment and the power required for this cooling

- Excessive power consumption by older, less energy-efficient technology

- Insufficient primary or standby power

- Rising energy costs and insufficient availability of power

- Lack of sufficient floor space to support growth and use of heavier and denser equipment

- Aging and limited HVAC (heating, ventilating, and air conditioning) capabilities

- Disposing of older technology in compliance with recycling regulations

- Complying with environmental health and safety mandates

- Improving infrastructure and application service delivery and enhancing productivity

- Doing more with less—less budget, less head count, and more IT equipment to support

- Support applications and changing workloads with adaptive capabilities

The available supply of electricity is being impacted by aging and limited generating and transmission capabilities as well as rising fuel costs. While industries such as manufacturing consume ever more electrical power, IT data centers and the IT equipment housed in those habitats require continued and reliable power.

IT data centers rely on available power and transmission capabilities, which are being affected by rising fuel costs and increasing demands as shown in Figure 2.1. The U.S. Environmental Protection Agency (EPA) estimates that with no changes, U.S. IT data center electric power consumption will jump to 3% of the U.S. total by 2010–2012. IT data centers require ever more power, cooling, and physical floor space to accommodate the servers, storage, and network components necessary to support growing application demands. In an era of growing environmental awareness, IT data centers, information factories of all sizes, and enterprise data centers in particular have issues and challenges pertaining to

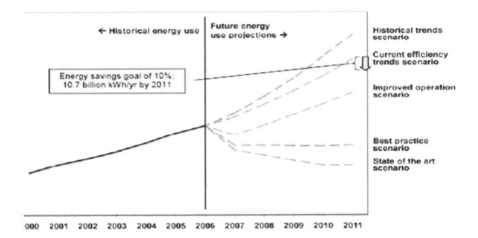

Figure 2.1   Projected Electricity Use, 2007–2011 (Source: U.S. Environmental Protection Agency)

power, cooling, floor space, and greenhouse gas emissions as well as "clean" disposal of IT equipment.

Data center demand for electrical power is also in competition with other power consumers, leading to shortages and outages during peak usage periods. There are also increasing physical requirements for growing data centers in the form of more servers, storage, and network components to support more IT and related services for business needs. Other pressing issues for IT data centers are cooling and floor space to support more performance and storage capacity without compromising availability and data protection.

Thus, if a data center is at its limit of power, and if the data center needs to increase processing and storage capabilities by 10% per year, a corresponding improvement in efficiency of at least the same amount is required. Over the past decade or so, capacity planning has been eliminated in many organizations because of the lowering cost of hardware; however, there is an opportunity to resurrect the art and science of capacity planning to tie power and cooling needs with hardware growth and to implement data center power demand-side management to ensure sustained growth.

Figure 2.2 shows typical power consumption and energy usage of typical data center components. With a current focus on boosting performance

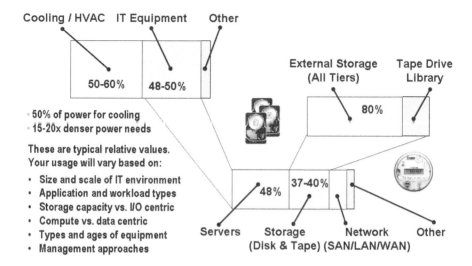

Figure 2.2   Average IT Data PCFE Consußmption (Source: www.storageio.com)

and reducing power consumption for servers and their subsequent cooling requirements, the PCFE focus will shift to storage. Even with denser equipment that can do more work and store more information in a given footprint, continued demand for more computing, networking, and storage capability will keep pressure on available PCFE resources. Consequently, addressing PCFE issues will remain an ongoing issue, and, thus, performance and capacity considerations for servers, storage, and networks need to include PCFE aspects and vice versa.

2.2   Electrical Power—Supply and Demand Distribution

Adequate electrical power is often cited as an important IT data center issue. The reasons for lack of available electrical power can vary greatly. Like data networks, electrical power transmission networks, also known as the power grid, can bottleneck. For example, there may be sufficient power or generating capabilities in your area, but transmission and substation bottlenecks may prevent available power from getting where it is needed. For example, consider the following scenarios:

- Power is available to the facility, but the facility's power distribution infrastructure is constrained.
- Power is available to the facility, but standby or backup power is insufficient for growth.
- Power is available in the general area, but utility constraints prevent delivery to the facility.
- Power costs are excessive in the region in which the IT equipment and facilities are located.

General factors that affect PCFE resource consumption include:

- Performance, availability, capacity, and energy efficiency (PACE) of IT resources
- Efficiency of HVAC and power distribution technologies
- The general age of the equipment—older items are usually less efficient
- The balance between high resource utilization and required response time
- Number and type of servers, type of storage, and disk drives being used
- Server and storage configuration to meet PACE service-level requirements

There is a correlation between how IT organizations balance server, storage, and networking resources with performance and capacity planning and what electrical **generating and transmission (G&T)** utilities do. That correlation is capacity and demand management. For G&T utilities, building or expanding existing generating and transmission facilities are cost- and time-consuming activities, so G&Ts need to manage the efficient use of resources just as IT managers must. G&Ts rely on supervisory control and data acquisition management systems to collect data on G&T components and to enable real-time management of these resources.

Thus, there are many similarities between how IT centers manage resources using simple network management protocol traps and alerts together with capacity planning and how the G&T industry manages its

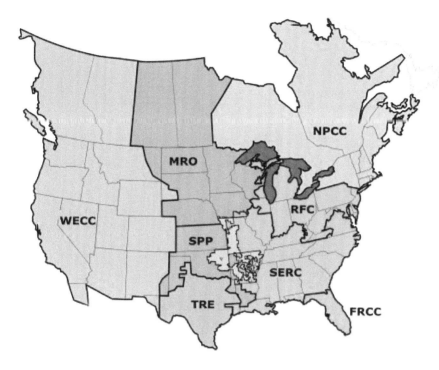

Figure 2.3   North American Electrical Power Grid (Source: www.nerc.com)

resources. Think of the power plants as servers and the transmission grid as a network. Like IT data centers, which have historically used performance and capacity planning to maximize their resources versus the expense of buying new technologies, the power companies do so on an even larger scale. Power companies have a finite ceiling to the amount of power they can provide, and data centers have a ceiling on the amount of power they can consume based on available supply.

Figure 2.3 shows how electrical power is managed in different regions of the United States as part of the North American electrical power grid. The U.S. power grid is, as is the case in other parts of the world, a network of different G&T facilities that is able to shift available power to different areas within a region as needed.

Figure 2.4 shows a simplified example of how the G&T power grid works, including generation facilities (power plants), high-voltage distribution lines, and lower-voltage distribution from local substations. IT data centers typically receive electric power via local distribution from one or more substation power feeds. As in an IT data network, there are many

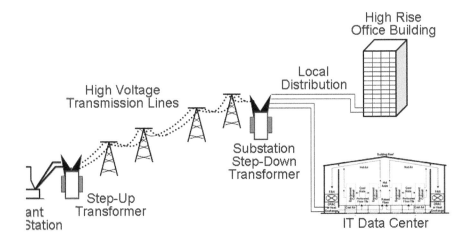

Figure 2.4   Electrical Power G&T Distribution

points for possible bottlenecks and contention within the G&T infrastructure. As a result, additional power may not be available at a secondary customer location; however, there may be power available at a primary substation or at a different substation on the G&T grid.

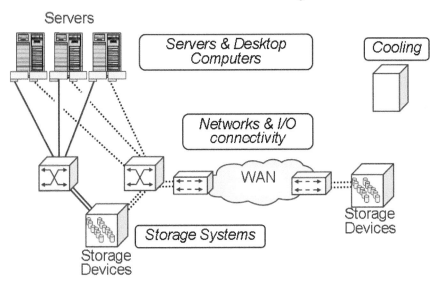

Figure 2.5   IT Data Center Consumers of Power

Once electricity is supplied to an IT data center, various devices, some of which are shown in Figure 2.5, consume the power. Additional items not shown in Figure 2.5 that consume power include HVAC or computer room air conditioning, power conversion, and distribution, along with general facility components such as battery chargers and lights.

2.3   Determining Your Energy Usage

When was the last time you looked at your business energy bill or your home electric bill? If you have not done so recently, look at or ask someone about what your facilities energy bill looks like. Also, look at your home energy bill and see how many kilowatt-hours you used, what the base energy rates are, what surcharges and other fees are assessed, and other information. Once you have reviewed your energy bill, can you determine where, when, and how electrical power is being used?

There are different approaches to determining energy usage. One is to take a macro view, looking at how much energy a data center consumes in total and working down to individual devices. The opposite approach is to work backwards from a micro view of components in a device and total the results across the various devices. Measured electricity usage can be obtained from utility bills, meters, or facilities management software tools, depending on what is available. Sources for estimating power consumption on a device or component level are vendor-supplied information, including equipment faceplates, and site planning guides. Other sources for determining energy usage are power meters and analyzers as well as power distribution devices that can also measure and report on power usage.

Electrical power is typically charged by utilities based on the number of kilowatt-hours used or the number of 1,000 Watt (W) of electricity used. For example, a server that draws 500 W consumes 0.5 kWh, or a storage device that consumes 3,500 W when being actively used consumes 3.5 kWh. Note that while energy usage varies over time and is cumulative, energy from a utility billing standpoint is based on what is used as of an hour. That is, if a server draws 500 W, its hourly energy bill will be 500 W or 0.5 kWh, as opposed to 500 × 60 seconds × 60 minutes. Likewise, electrical power generation is quoted in terms of kilowatt-hours or mega (million) watt-hours (MWh) as of a given point in time. For example, an 11-MW power plant is capable of producing 11,000 kWh at a given point in time, and if usage is constant, the energy is billed as 11,000 kWh.

Typically, energy usage is based on metered readings, either someone from the utility company physically reading the meter, remote reading of the meter, or, perhaps, estimated usage based on historical usage patterns, or some combination. Electric power is charged at a base rate (which may vary by location, supply, and demand fuel sources, among other factors) per kilowatt-hour plus fuel surcharges, peak demand usage surcharges, special fees, applicable commercial volume peak usage minus any applicable energy saver discounts. For example, for voluntarily reducing power consumption or switching to standby power generation during peak usage periods, utilities may offer incentives, rebates, or discounts.

IT technology manufacturers provide information about electrical energy consumption and/or heat (Btu per hour) generated under a given scenario. Some vendors provide more information, including worst-case and best-case consumption information, while others provide only basic maximum breaker size information. Examples of metrics published by vendors and that should be visible on equipment include kilowatts, kilovolts, amperage, volts AC or Btu. Chapter 5 discusses various PCFE-related metrics and measurements, including how to convert from watts to Btu and from amperes to watts.

To calculate simple energy usage, use the values in Table 2.1, selecting the energy costs for your location and the number of kilowatt-hours required to power the device for one hour. For example, if a server or storage device consumes 100 kWh of power and the average energy cost is 8¢/kWh, energy cost is $70,100 annually. As another example, a base rate for 1 kWh might be 12¢/kWh but 20¢/kWh for usage over 1,000 kWh per month. Note that this simple model does not take into consideration regional differences in cost, demand, or availability, nor does it include surcharges, peak demand differentials, or other factors. The model also does not differentiate between energy usage for IT equipment operation and power required for cooling. The annual kWh is calculated as the number of kWh × 24 × 365.

A more thorough analysis can be done in conjunction with a vendor environment assessment service, with a consultant, or with your energy provider. As a start, if you are not in the habit of reading your monthly home energy bill, look to see how much energy you use and the associated costs, including surcharges and fees.

Table 2.1   Example Annual Costs for Various Levels of Energy Consumption

Hourly Power Use (kWh)	5¢/kWh	8¢/kWh	10¢/kWh	12¢/kWh	15¢/kWh	20¢/kWh
1 kWh	$438	$701	$806	$1,051	$1,314	$1,732
10 kWh	$4,380	$7,010	$8,060	$10,510	$13,140	$17,520
100 kWh	$43,800	$70,100	$80,600	$105,100	$131,400	$175,200

2.4   From Energy Avoidance to Efficiency

Given specific goals, requirements, or objectives, shifting to an energy-efficient model can either reduce costs or enable new IT resources to be installed within an existing PCFE footprint. Cost reductions can be in the form of reducing the number of new servers and associated power and cooling costs. An enabling growth and productivity example is to increase the performance and capacity, or the ability to do more work faster and store more information in the same PCFE footprint. Depending on current or anticipated future power and/or cooling challenges, several approaches can be used to maximize what is currently in place for short-term or possibly even long-term relief. Three general approaches are usually applied to meet the various objectives of data center power, cooling, floor space, and environmental aims:

- Improve power usage via energy efficiency or power avoidance.
- Maximize the use of current power—do more with already available resources.
- Add additional power, build new facilities, and shift application workload.

Other approaches can also be used or combined with short-term solutions to enable longer-term relief, including:

- Establish new facilities or obtain additional power and cooling capacity.

- Apply technology refresh and automated provisioning tools.
- Use virtualization to consolidate servers and storage, including thin provisioning.
- Assess and enhance HVAC, cooling, and general facility requirements.
- Reduce your data footprint using archiving, real-time compression and de-duplication.
- Follow best practices for storage and data management, including reducing data sprawl.
- Leverage intelligent power management such as MAID 2.0-enabled data storage.
- Use servers with adaptive power management and 80% Plus efficient power supplies.

Virtualization is a popular means of consolidating and eliminating underutilized servers and storage to reduce cost, electricity consumption, and cooling requirements. In their place, power-efficient and enhanced-performance servers and storage, including blade centers, are being deployed to support consolidated workloads; this is similar to what has historically been done in enterprise environments with IBM mainframe systems. However, for a variety of reasons, not all servers, storage, or networking devices lend themselves to being consolidated.

Some servers and storage as well as network devices need to be kept separate to isolate different clients or customers, different applications or types of data, development and test from production, online customer-facing systems from back-end office systems, or for political and financial reasons. For example, if a certain group or department bought an application and the associated hardware, that may prevent those items from being consolidated. Department turf wars can also preclude servers and storage from being consolidated.

Two other factors that can impede consolidation are security and performance. Security can be tied to the examples previously given, while application performance and size can have requirements that conflict with those of applications and servers being consolidated. Typically, servers with applications that do not fully utilize a server are candidates for consolidation. However, applications that are growing beyond the limits of a single dual-, quad-, or multi-core processor or even cluster of servers do not lend themselves to consolidation. Instead, this latter category of servers and applications need to scale up and out to support growth.

Industry estimates and consensus vary from as low as 15% to over 85% in terms of actual typical storage space allocation and usage for open systems or non-mainframe-based storage, depending on environment application, storage systems, and customer service-level requirements. Low storage space capacity usage is typically the result of one or more factors, including the need to maintain a given level of performance to avoid performance bottlenecks, over-allocation to support dynamic data growth, and sparse data placement because of the need to isolate applications, users, or customers from each other on the same storage device. Limited or no insight as to where and how storage is being used, not knowing where orphaned or unallocated storage is stranded, and buying storage based on low cost per capacity also contribute to low storage space capacity utilization.

The next phase of server virtualization will be to enhance productivity and application agility in order to scale on a massive basis. Combined with clustering and other technologies, server virtualization is evolving to support scaling beyond the limits of a single server—the opposite of the server consolidation value proposition. Similarly, server virtualization is also extending to the desktop to facilitate productivity and ease of management. In both of these latter cases, transparency, emulation, and abstraction for improved management and productivity are the benefits of virtualization.

2.5   Energy Efficiency Incentives, Rebates, and Alternative Energy Sources

Carbon offsets and emissions taxes have their place, particularly in regions where legislation or regulations require meeting certain footprints. In such locations, a business decision can be to do an analysis of paying the emissions tax fee to comply near term versus cost to comply long term. In other words, pay carbon offsets or get money back and achieve efficiency.

Some U.S. energy utilities provide incentives and rebates for energy efficiency and/or use of renewable energy. The programs vary by utility, with some being more advanced than others, some more defined, and some more customer oriented. Some programs provide rebates or energy savings, while others provide grants or funding to conduct energy efficiency assessments or

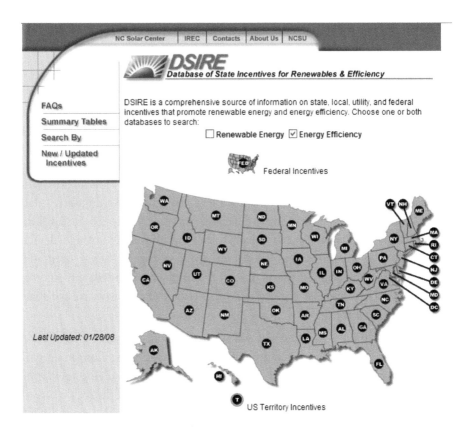

Figure 2.6   Database of State Incentives for Renewables & Efficiency (DSIRE)

make infrastructure and facilities changes. In general, utilities do not differentiate between an IT data center and vendor development and testing lab facilities or consider the size of the data center.

Pacific Gas and Electric (PG&E) has been a pioneer in energy rebates incentives for IT data centers. PG&E has programs targeted toward energy demand-side management for various localities and industry sectors. Other energy utilities also leverage demand-side management as part of their capacity planning and performance management of their resources: energy generation via their G&T facilities. A group of utilities led by PG&E has created a Consortium for Energy Efficiency (CEE) to exchange and coordinate ideas and to further develop specific programs to address IT data center power issues. Another venue is DSIRE (Database of State Incentives for Renewables & Efficiency), which is a portal that provides information on various available energy efficiency as well as

Figure 2.7   The EPA Power Portal (Source: U.S. Environmental Protection Agency)

renewable energy programs across the United States. As shown in Figure 2.6, additional information about such incentives on a state-by-state basis can be found at ww.dsireusa.org.

The EPA has many programs associated with power and energy that combine an environmental viewpoint with a perspective on sustaining supply to meet demand. Examples of EPA programs include Energy Star, Green Power (www.epa.gov/grnpower), and others. The EPA Green Power portal shown in Figure 2.7 provides information on various programs, including alternative and green power sources, on a state-by-state basis. Other agencies in different countries also have programs and sources of information, for example, the Department for Environment, Food and Rural Affairs (DEFRA; www.defra.gov.uk), in the United Kingdom. In Canada, Bullfrog Power has a portal (www.bullfrogpower.com) that provides information on green and alternative power for homes and businesses.

Fossil fuels for primary and secondary electric power generation are coal, oil and gas (natural gas, liquefied propane [LP] gas, gasoline or aviation fuel, diesel). Alternative and renewable sources for electricity generation include biomass (burning of waste material), geothermal, hydro, nuclear, solar, wave and tidal action, and wind. Part of creating an energy-efficient and environmentally friendly data center involves leveraging different energy sources for electricity. For example, a local power utility can

Table 2.2   Some Relevant Standards, Regulations, and Initiatives

Abbreviation or Acronym	Description
DOE FEMP	U.S. Department of Energy Federal Energy Management Program
ECCJ	Energy Conservation Center Japan (www.eccj.or.jp)
ELV	End of Life Vehicle Directive (European Union)
Energy Star	U.S. EPA Energy Star program (www.energystar.gov/data-centers)
EPEAT	Electronic Product Environmental Assessment Tool (www.epeat.net)
ISO 14001	Environmental management standards
JEDEC	Joint Electronic Device Engineering Council (www.jedec.org)
JEITA	Japan Electronics Information Technology Industry Association (www.jeita.or.jp)
J-MOSS	Japanese program for removal of hazardous substances
LEED	Leadership in Energy Efficiency Design
MSDS	Material Safety Data Sheet for products
NRDC	Natural Resources Defense Council (www.nrdc.org)
REACH	Registration, Evaluation, Authorization and Restriction of Chemicals
RoHS	Restriction of Hazardous Substances (www.rohsguide.com)
SB20/50	California Electronics Waste Recycling Act of 2003
USGBC	U.S. Green Building Council (www.usgbc.org)
WEEE	Waste from Electrical and Electronic Equipment
WGBC	World Green Building Council (www.worldgbc.org)

provide a primary source for electric power, leveraging the lowest-cost, most effectively available power currently available in the power grid. As a standby power source, backup generators fueled by diesel, propane, or LP gas can be used.

From an economic standpoint, working with local and regional utilities to improve electrical efficiency and obtain rebates and other incentives should all be considered. For example, during nonpeak hours, electrical power from the local power grid can be used; during peak demand periods, backup standby generators can be used in exchange for reduced energy fees and avoiding peak demand surcharges. Another economic consideration, however, is the cost to run on standby generator power sources. These costs, including fuel and generator wear and tear, should be analyzed with respect to peak-demand utility surcharges and any incentives for saving energy. For organizations that have surplus self-generated power, whether from solar, wind, or generators, some utilities or other organizations will buy excess power for distribution to others, providing cash, rebates, or discounts on regular energy consumption. Learn more about electrical energy fuel sources, usage, and related statics for the United States and other countries at the energy information administration website www.eia.doe.gov/emeu/aer/elect.html.

2.6   PCFE and Environmental Health and Safety Standards

The green supply chain consists of product design, manufacture, distribution, and retirement, along with energy production, deliver, and consumption. Table 2.2 provides a sampling of initiatives relating to PCFE and environmental health and safety.

Later chapters in this book will present additional technologies and techniques to boost efficiency and productivity to address PCFE issues while balancing performance, availability, capacity, and energy efficiency to meet various application service requirements.

2.7   Summary

Action items suggested in this chapter include:

- Gain insight into how electrical power is used to determine an energy efficiency baseline.
- Investigate rebates and incentives available from utilities and other sources.
- Explore incentives for conducting data center energy efficiency assessments.

- Understand where PCFE issues and bottlenecks exist and how to address them.
- Investigate alternative energy options, balancing economic and environmental concerns.
- Review your home and business electric utility bills to learn about power usage and costs.
- Learn more about the various regulations related to environmental health and safety.

Other takeaways from this chapter include:

- Energy avoidance may involve powering down equipment
- Energy efficiency equals more useful work and storing more data per unit of energy.
- Virtualization today is for consolidation
- Virtualization will be used tomorrow to enhance productivity.

Chapter 3

What Defines a Next-Generation and Virtual Data Center?

Virtual data centers enable data mobility, resiliency, and improved IT efficiency.

In this chapter you will learn:

- What defines a green and virtual data center
- How virtualization can be applied to servers, storage, and networks
- The many faces of server, storage, and networking virtualization
- How to leverage virtualization beyond server, storage, and network consolidation
- The various components and capabilities that comprise a virtual data center
- How existing data centers can transform into next-generation virtual data centers

Many approaches and technologies, addressing different issues and requirements, can be used to enable a green and virtual data center. Virtualization is a popular approach to consolidating underutilized IT resources, including servers, storage, and input/output (I/O) networks to free up floor-space, lower energy consumption, and reduce cooling demand, all of which can result in cost savings. However, virtualization—and particularly consolidation—applies to only a small percentage of all IT resources. The importance of this chapter is that there are many facets of virtualization that can be used to enable IT infrastructure resource management to improve service delivery in a more cost-effective and environmentally friendly manner.

In Chapters 1 and 2 we discussed various green and environmental issues. In this chapter we turn to what can be done and how to leverage different technologies to enable a green virtual data center. Here we look at what defines a green and virtual data center, building on the foundation laid out in Part I.

This chapter also looks at how virtualization can be applied to servers, storage, and networks. Virtualization can be applied to consolidate underutilized IT resources. It can also be used to support transparent management, enabling scaling for high growth and performance applications in conjunction with clustering and other technologies.

3.1   Why Virtualize a Data Center?

A virtual data center can, and should, be thought of as an information factory that needs to run 24-7, 365 days a year, to delivery a sustained stream of useful information. For the information factory to operate efficiently, it needs to be taken care of and seen as a key corporate asset. Seen as an asset, the IT factory can be invested in to maintain and enhance productivity and efficiency, rather than being considered a cost center or liability.

The primary focus of enabling virtualization technologies across different IT resources is to boost overall effectiveness while improving application service delivery (performance, availability, responsiveness, security) to sustain business growth in an economic and environmentally friendly manner. That is, most organizations do not have the luxury, time, or budget to deploy virtualization or other green-related technologies and techniques simply for environmental reasons—there has to be a business case or justification.

Virtual data centers (Figure 3.1), regardless of whether new or existing; require physical resources including servers, storage, and networking, and facilities to support a diverse and growing set of application capabilities while sustaining business growth. In addition to sustaining business growth, applications need to be continually enhanced to accommodate changing business rules and enhance service delivery. Application enhancements include ease of use, user interfaces, rich media (graphics and video, audio and intuitive help), along with capturing, storing, and processing more data.

There are a growing number of business cases and justifications for adopting green technologies that reduce costs or maximize use of existing resources while also benefiting the environment. However, it is rare for a business to have surplus budget dollars, personnel, facilities, and management support to deploy new technologies simply for the sake of deploying them.

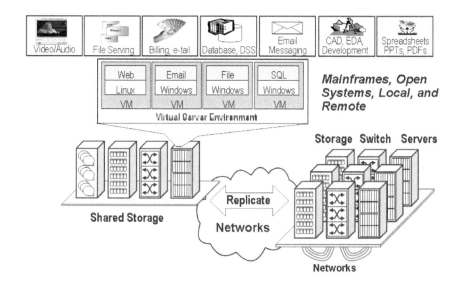

Figure 3.1   Virtual Data Center with Physical Resources

Challenges facing IT data centers and businesses of all sizes include:

- Ensuring reliability, availability, serviceability, and management
- Dealing with performance bottlenecks and changing workloads
- Securing applications and data from diverse internal and external threat risks
- Enabling new application features and extensibility while supporting business growth
- Managing more data for longer periods of time
- Supporting compliance or self-governance while securing intellectual property data
- Maximizing **power, cooling, floor space, and environmental (PCFE)** issues
- Ensuring data and transaction integrity for business continuance and disaster recovery
- Consolidating IT resources to reduce complexity and contain management costs

- Supporting interoperability with existing applications, servers, storage, and networks

- Reducing or abstracting complexities associated with applications interdependencies

- Leveraging existing IT personal skill sets and experience

- Scaling existing and new applications with stability and in cost-effectively

- Moving data transparently from old to new storage during technology replacement

- Enhancing IT service delivery in a cost-effective and environmentally friendly manner

IT data centers are increasingly looking to virtualization technology and techniques to address all these issues and activities. Virtualization can be used to consolidate servers and storage resources in many environments to boost resource utilization and contain costs—for example, using a software based **virtual machine (VM)** to combine applications and operating systems images from underutilized physical servers to virtual machines on a single server. With a growing focus on PCFE and the greening of IT in general, consolidating servers, as well as storage, networking, and facilities, is a popular and easy-to-understand use of virtualization.

Energy efficiency today can sometimes mean simply energy avoidance. In the future, however, emphasis will shift to doing and enabling more work and storing increasing amounts of information for longer periods of time. This will have to be accomplished while consuming less energy and using less floor space. Doing more work in a more productive manner with less energy will result in efficiencies from improved technologies, techniques and best practices.

Consequently, a green and virtual data center should be much more than just an environment that leverages some virtualization for consolidation purposes. A green and virtual data center should enable transparent management of different physical resources to support flexible IT service delivery in an environmentally and economically friendly manner on both local and remote bases.

Scaling with stability means that as performance is increased, application availability or capacity is neither decreased nor is additional management complexity or cost introduced. Similarly, scaling with stability means that as capacity is increased, neither performance nor availability suffer, nor is performance negatively affected by growth or increased workload or application functionality. This also includes eliminating single points of failure and supporting fault isolation and containment, self-healing, supporting mixed performance of small and large I/O operations, and additional functionality or intelligence in technology solutions without adding cost or complexity. In addition, scaling with stability means not introducing downtime or loss of availability or negative performance as a result of growth. High resilience or self-healing with fault isolation and containment will prevent single points of failure from cascading into rolling disasters.

Flexibility and agility will enable virtual data centers to meet changing business and application requirements while quickly and transparently adopting new technology enhancements without disruption.

Tenets of a green and virtualized data center include:

- Flexible, scalable, stable, agile, and highly resilient or self-healing systems
- Quick adaptation and leverage of technology improvements
- Transparency of applications and data from physical resources
- Efficient operation without loss of performance or increased cost complexity
- Environmentally friendly and energy efficient yet economical to maintain
- Highly automated and seen as information factories as opposed to cost centers
- Measureable with metrics and reporting to gauge relative effectiveness
- Secure from various threat risks without impeding productivity

3.2 Virtualization Beyond Consolidation—Enabling Transparency

There are many facets of virtualization. Aggregation is a popular approach to consolidate underutilized IT resources including servers, storage, and networks. The benefits of consolidation include improved efficiency by eliminating underutilized servers or storage to reduce electrical power, cooling, and floor space requirements as well as management time, or to reuse and repurpose servers that have become surplus to enable growth or support new application capabilities.

Another form of virtualization is emulation or transparency providing abstraction to support integration and interoperability with new technologies while preserving existing technology investments and not disrupting software procedures and policies. Virtual tape libraries are a commonly deployed example of storage technology that combines emulation of existing tape drives and tape libraries with disk-based technologies. The value proposition of virtual tape and disk libraries is to coexist with existing backup software and procedures while enabling new technology to be introduced.

Figure 3.2 shows two examples of virtualization being used, with consolidation on the left side and transparency for emulation and abstraction to support scaling on the right. On the consolidation side, the operating systems and applications of multiple underutilized physical servers are consolidated onto a single or, for redundancy, multiple servers in a virtual environment with a separate virtual machine emulating a physical machine. In this example, each of the operating systems and applications that were previously running on their own dedicated servers now run on a virtual server. Consolidation enables multiple underutilized servers to be combined yet let each system think and operate as though it still had its own server.

For a variety of reasons, not all servers or other IT resources lend themselves to consolidation. These reasons may include performance, politics, finances, service-level, or security issues. For example, an application may need to run on a server with low CPU utilization to meet performance and response-time objectives or to support seasonal workload adjustments. Also, certain applications, data, or even users of servers may need to be isolated from each other for security and privacy reasons.

Politics and financial, legal, or regulatory requirements also need to be considered. For example, a server and application may be "owned" by

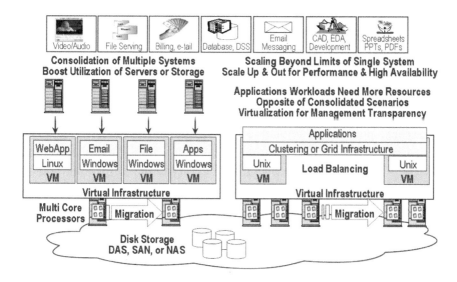

Figure 3.2   Consolidation vs. Transparent Management with Virtualization

different departments or groups and thus managed and maintained separately. Regulatory or legal requirements may dictate that certain systems be kept separate from other general-purpose or mainstream applications, servers, and storage. Separation of applications may also be necessary isolate development, test, quality assurance, back-office, or other functions from production or online applications and systems, as well as to support business continuance, disaster recovery, and security.

For applications and data that do not lend themselves to consolidation, a different way to use virtualization is to enable transparency of physical resources to support interoperability and coexistence between new and existing software tools, servers, storage, and networking technologies, such as enabling new, more energy-efficient servers or storage with improved performance to coexist with existing resources and applications.

On the right side in Figure 3.2 are examples of applications that don't lend themselves to consolidation because they need to be isolated from other applications or clients for performance or other reasons. However, these applications can still benefit from transparency and abstraction when combined with clustering technology to enable scaling beyond the limits of a single server, storage, or networking device. Included on the right side of Figure 3.2 are applications that need more resources for performance or availability than are available from a single large server and hence need to scale horizontally (also known as scale up and out).

Another use of virtualization transparency is to enable new technologies to be moved into and out of running or active production environments to facilitate technology upgrades and replacements. Still another use is to adjust physical resources to changing application demands, such as seasonal planned or unplanned workload increases. Transparency via virtualization also enables routine planned and unplanned maintenance functions to be performed on IT resources without disrupting applications and users of IT services.

Virtualization in the form of transparency or abstraction of physical resources to applications can also be used to help achieve energy savings and address other green issues by enabling newer, more efficient technologies to be adopted more quickly. Transparency can also be used to implement tiered servers and storage to leverage the right technology and resource for the task at hand as of a particular point in time.

Business continuity (BC) and **disaster recovery (DR)** are other areas in which transparency via virtualization can be applied to in a timely and cost-efficient manner in-house, via a managed service provider, or using some combination. For example, traditionally, a BC or DR plan requires the availability of similar server hardware at a secondary site. Some challenges with this kind of redundancy are that the service and servers must be available to those who need them when they need them. For planned testing, this may not be a problem; however, in the event of a real disaster, a first-come, first-served situation could arise, with too many subscribers to the same finite set of physical servers, storage, and networking facilities.

If dedicated and guaranteed servers and storage resources are available for BC and DR, competition for resources is eliminated. This means, however, that additional servers and storage need to be powered, cooled, and given floor space and management. In addition, these operating systems and applications may require identical or very similar hardware and configurations.

3.3   Components of a Virtual Data Center

For some organizations, the opportunity to start from scratch with a new green data center may exist. However, for most, enabling a virtualized data center relies on transforming existing facilities, servers, storage, networking, and software tools along with processes and procedures to adopt

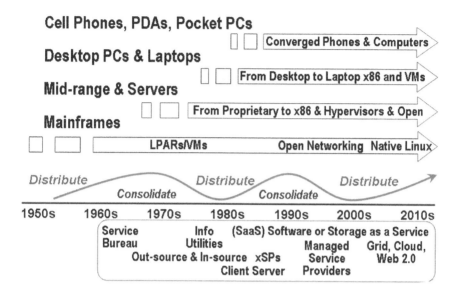

Figure 3.3   IT Resource Computing and Consolidation Continuum

virtualization technologies and techniques. Thus, virtualization technologies should enable existing IT resources to be transformed and transition to a next-generation virtualization data center environment. The benefit is support of growth and enhancement of service delivery in a cost-effective and timely manner.

A green and virtual data center is more than an environment that leverages server, storage, or network virtualization to improve resource usage. IT resource consolidation has been a recurring theme over the past several decades with the shift from centralized to distributed computing. The cycle shown in Figure 3.3 went from distributed resources to consolidation, followed by client–server systems, followed by reconsolidation, followed by Internet dispersion of resources to the current consolidation phase.

Some reasons for the cycle of distribute and consolidate include changing business and IT models, technology trends, and financing models. Another factor is the decades-old issue of addressing the server-to-storage I/O performance gap, where the cost of hardware continues to decrease and servers becoming faster and smaller. Meanwhile, storage capacity and availability continue to increase while physical footprint and price decrease. However, there is a gap between storage capacity, server, and storage performance (see Figure 3.4). The result is that for some

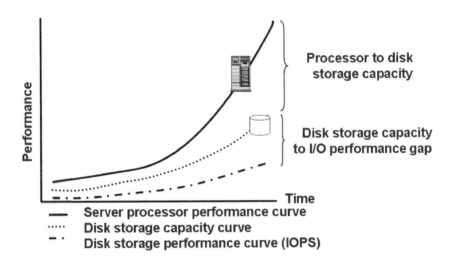

Figure 3.4   Server and Storage I/O performance Gap

applications, to achieve a given level of performance, more resources (disks, controllers, adapters, and processors) are needed, resulting in a surplus of storage capacity.

In an attempt to reduce the excess storage capacity, consolidation sometimes happens without an eye on performance, looking only at the floor space, power, and cooling benefits of highly utilized storage. Then, to address storage performance bottlenecks, the storage gets reallocated across more storage systems, and the cycle starts again. Another driver for underutilized hardware is a result of servers and storage being bought by individual departments or for specific applications, with politics and financial constraints limiting their shared usage.

Another variation is the notion that hardware is inexpensive, so buy more. With higher energy prices, simply throwing more hardware at application performance or other issues is no longer an option for environments with PCFE constraints.

An opportunity enabled by virtualization transparency and abstraction is to address performance and related issues with tiered servers, storage, and networks using high-performance, energy-efficient devices balanced with high-capacity energy-efficient devices. For example, instead of using several low-cost disk drives and associated adapters for performance-intensive applications, use faster storage measured on an activity-per-watt (IOPS/

watt, bandwidth/watt, transaction or message/watt) basis instead of on a cost-per-gigabyte or energy-per-gigabyte basis.

3.3.1   Infrastructure Resource Management Software Tools

Falling under the umbrella of **infrastructure resource management (IRM)** are various activities, tools, and processes for managing IT resources across different technology domains (servers, storage, networks, facilities, and software) with diverse interdependencies to enable IT application service delivery.

Aspects of IRM include:

- Logical and physical security, including rights management and encryption
- Asset management, including configuration management databases
- Change control and management, along with configuration validation management tools
- Data protection management, including business continuity and disaster recovery
- Performance and capacity planning and management tools
- Data search and classification tools for structured and unstructured data
- High-availability and automated self-healing infrastructures
- Data footprint reduction, including archiving, compression, and de-duplication
- Planning and analysis, event correlation, and diagnostics
- Provisioning and allocation of resources across technology domains
- Policies and procedures, including best practices and usage template models

3.3.2   Measurements and Management Insight

Various metrics and measurements are needed in order to provide insight into how data centers and applications are running as well as using resources. Metrics and measurements are also important for timely and proactive problem resolution and isolation, as well as event correlation to support planning and reconfiguration for improved service delivery and growth. Metrics and management insight are also needed to ensure compliance and other requirements are being met, including security or activity logs, as well as that data is being protected as it is intended and required to be.

Examples of metrics, measurements, and reporting include:

- Energy consumption and effectiveness of work being performed
- Server, storage, and network performance and capacity usage information
- Availability of IT resources, including planned and unplanned downtime
- Effectiveness of IT resources to meet application service-level objectives
- Data protection management status and activity
- Error, activity and events logs, data protection status, and alarms
- Metrics for recycling, carbon disclosure, and environmental health and safety reporting

Measurements and monitoring of IT resources are key to achieving increased efficiency so that the right decisions can be made for the right reasons while addressing and fixing problems rather than simply moving them around. For example, if applications require overallocation of server, storage, and networking resources to meet performance and application service objectives, consider options such as leveraging faster technologies that consume less power to accomplish the necessary work.

3.3.3   Facilities and Habitats for Technology

One potentially confusing aspect of next-generation data centers is the implication that they must be built from scratch, as new facilities with all new technology, IT equipment, and software. For some environments, that may be the case. For most environments, however, even if a new physical facility is being built or an existing one expanded or remodeled, integration with existing technologies and management tools is required. Consequently, the road to a virtual or next-generation and green data center is an evolution from a current environment to a new and enhanced way of operating and managing IT resources in an efficient and flexible manner. This means, among other things, that existing legacy mainframes may need to coexist with current-generation blade servers or other modern servers. Similarly, magnetic tape devices may need to coexist with newer disk-based systems, and LANs may need to coexist with networks using copper, optical, and even wireless communications.

All of these IT resources need to be housed in a technology-friendly environment that is or will become more energy efficient, including in how cooling is handled as well as primary power distribution and provisions for standby power. (Chapter 6 looks more closely at various options for improving the energy efficiency of facilities.)

3.3.4   Tiered Servers and Software

Servers have received a lot of attention as prime consumers of electrical power and producers of heat. Consequently, virtualization in the form of server consolidation to combine multiple lower-utilized servers onto a single or fewer physical servers running virtual machines is a popular topic. Having fewer servers means that less electrical power is required for both the servers and the necessary cooling.

Many servers are underutilized at various times, and some are always underutilized as a result of how or when they were acquired and deployed. Through much of the late 1990s, the notion was that hardware was inexpensive, so the easiest solution seemed to be to throw hardware at various challenges as they arose. Or, as new applications came online, it may have been faster to acquire a new server than to try and find space on another server.

Given today's rising energy costs, concerns about PCFE issues, and the need to boost IT resource efficiency, many smaller or underutilized servers are being consolidated onto either larger servers or blade servers running virtual infrastructure software and virtual machines such as those from VMware, Virtual Iron, or Microsoft, among others. Not all servers and applications, however, lend themselves to consolidation, for the reasons we have discussed. As a result, some applications need to scale beyond the limits of a single device, where virtualization, as discussed earlier, enables transparency for maintenance, upgrades, load balance, and other activities.

3.3.5   Tiered Storage and Storage Management

There are many different ways of implementing storage virtualization, including solutions that aggregate heterogeneous or different vendors' storage to enable pooling of resources for consolidated management. Although it is popular to talk about, storage aggregation has trailed in actual customer deployments to other forms of virtualization such as emulation.

Virtual tape libraries (VTLs) or virtual tape systems (VTS) or disk libraries and de-duplication appliances that emulate the functionality of previous-generation storage solutions are the most common forms of storage virtualization in use today. The benefit of emulation is that it enables abstraction and transparency as well as interoperability between old processes, procedures, software, or hardware and newer, perhaps more energy-efficient or performance-enhancing, technologies.

The next wave of storage virtualization looks to be in step with the next server virtualization wave—virtualization not for consolidation or emulation but to support transparent movement of data and applications over tiered storage for both routine and non-routine IRM functions. These systems will facilitate faster and less disruptive technology upgrades and expansion so that storage resources can be used more effectively and transparently. Moving forward, there should be a blurring of the lines between transparency and abstraction vs. consolidation and aggregation of resources.

Another form of virtualization is partitioning or isolation of consolidated and shared resources. For example, some storage devices enable logical unit numbers (LUNs) to be mapped into partitions or groups to abstract LUN and volume mapping for coexistence with different servers. Another variation enables a storage server to be divided up into multiple logical virtual filers to isolate data from different applications and customers.

3.3.6   Tiered Networks and I/O Virtualization

There are many different types of networks, and convergence may include virtual connect infrastructures inside blade center servers, top-of-rack and end-of-rack solutions, modular switches and routers, as well as core and backbone directors for traditional networks as well as converged virtual I/O and I/O virtualization networks. There are also many kinds of networks and storage interfaces for connecting physical and virtual servers and storage.

3.3.7   Virtual Offices, Desktops, and Workstations

Another component of a virtual data center environment is the remote and virtual or home office. Desktop and workstation virtualization is a natural extension of what is taking place with servers, storage, and networks to boost utilization and effectiveness as well as to address complexities in configuring and deploying large numbers of workstations and desktops while enabling virtual offices to access and use data when and where needed in a secure and flexible manner.

3.4   Summary

Green and next-generation virtual data centers should be highly efficient, flexible, resilient, and environmentally friendly while economical to operate. Current focus is on virtualization from a consolidation perspective, but in the future there will be even more opportunities for IT environments to adapt their processes, techniques, and technologies to sustain business growth and enhance application service delivery experience while reducing costs without compromising performance, availability, or ability to store and process more information. There are many aspects of data storage virtualization that address routine IT management and support tasks, including data protection, maintenance, and load-balancing for seasonal and transient project-oriented application workloads.

The vendor who controls and manages the virtualization software, whether on a server, storage, or in the network, controls the vendor lock-in or "stickiness." If you are looking to virtualization to eliminate vendor lock-in, it is important to make sure you understand what lock-in is being left and what lock-in will be in its place. Vendor lock-in is not a bad thing if the capabilities, efficiency, economics, and stability offered by a solution outweigh any real or perceived risks or issues. Bad technology and tools in the wrong hands for the wrong tasks make for a bad solution, while good technology and tools in the wrong hands for the wrong tasks make for a not-so-good solution. The goal is to put good tools and techniques in the right hands for the right tasks to make an enabling and good solution.

The idea is to leverage virtualization technologies in the form of abstraction and transparency or emulation combined with tiered servers, tiered storage, and tiered networks to align the right technology to the task at hand at a particular time. Start to fix the problems instead of moving them around or bouncing from distributed to consolidated and moving the distributed problems back to a main site. You can get management of an increasing amount of data and resources under control, and you can do more work with less energy while supporting growth.

Chapter 4

IT Infrastructure Resource Management

You can't go forward if you cannot go back.—Jerry Graham, IT professional

In this chapter you will learn:

- What infrastructure resource management (IRM) is
- Why IRM is an important part of a virtual data center
- How IRM differs from data and information life-cycle management
- Data protection options for virtualized environments

Leveraging various tools, technologies, and techniques to address various pain points and business challenges is key to enabling a green and virtual data center. Best practices, people, processes, and procedures combine with technology tools, hardware, software, networks, services, and facilities to enable a virtual data center. The importance of this chapter is to understand how all these elements coupled with existing and emerging technologies can be applied to improve IT service delivery in a cost-effective and energy-efficient manner. All of this together allows the data center to meet service-level requirements while sustaining business growth.

Tenets of a green and environmentally friendly virtual data center include improved productivity to enable more work to be processed, more information to be stored and accessed, while using less energy to boost productivity and business agility. A green and virtualized data center is:

- Flexible, scalable, stable, agile, and highly resilient or self-healing
- Able to adapt and leverage technology improvements quickly
- Application and data transparent from physical resources

- Efficient and effective without loss of performance or increased cost complexity

- Environmentally friendly and energy efficient yet economical to maintain

- Highly automated and seen as an information factories rather than a cost center

- Measurable with metrics and reporting to gauge relative effectiveness

- Secure from various threat risks without impeding productivity

Infrastructure resource management (IRM) is the collective term that describes the best practices, processes, procedures, and technology tools to manage IT data center resources. IRM has a focus across multiple technology domains (applications, servers, networking, storage, and facilities) to address effective and maximum resource usage to deliver a given level of application service or functionality. IRM focuses on processes, procedures, hardware, and software tools that facilitate application and data management tasks. Although there can be areas of overlap, the aim of IRM is to deliver application services and information to meet business service requirement objectives while addressing performance, availability, capacity, and energy (PACE) and power, cooling, floor space, and environmental (PCFE) requirements in a cost-effective manner. Examples of IRM encompassing functions and IT disciplines across different technology domains are shown in Figure 4.1.

Part of the process of implementing a virtual data center is to remove barriers and change traditional thinking such as hardware vs. software, servers vs. storage, storage vs. networking, applications vs. operating systems and IT equipment vs. facilities. A reality is that hardware cannot exist without software, and software cannot exist or function without hardware. Servers need networks; networks need storage. Collectively, all these IT resources need a habitat with adequate PCFE capabilities to be functional. As a result, a virtual data center looks at bridging gaps between different functional groups and technology domain areas to improve management and agility to support growth and improve application service delivery.

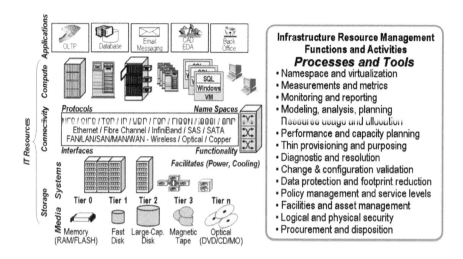

Figure 4.1   IRM Functions across Different Technology Domains

4.1   Common IRM Activities

As shown in Figure 4.1, there are many different tasks and activities along with various tools to facilitate managing IT resources across different technology domains. In a virtual data center, many of these tools and technologies take on increased interdependencies because the reliance on abstracting physical resources to applications and IT services.

Consequently, while specific technology domains may focus on specific areas, interdependencies across IT resource areas are a matter of fact for efficient virtual data centers. For example, provisioning a virtual server requires configuration and security of the virtual environment, physical servers, storage, and networks as well as associated software and facility-related resources. Backing up or protecting data for an application may involve multiple servers running different portions of an application requiring coordination of servers, storage, networks, software, and data protection tasks.

Common IRM activities involved with provisioning and managing IT resources include change control and configuration management, such as updating business continuity and disaster recovery (BC/DR) plans and documents or validatng configuration settings to avoid errors. Other tasks related to change control and configuration management include notification of changes and interdependencies to various systems or IT resources, fallback and contingency planning, maintaining templates, blueprints, run-books, and guides for configuration, as well as change and configuration testing and coverage analysis. Another IRM task involves configuring physical resources, including server and operating system or virtual machine setup, networking and input/output (I/O) configuration, storage system RAID (Redundant Array of Independent Disks), data protection and security, along with storage allocation including logical unit number (LUN) creation, mapping, and masking. Other configuration and resource allocation IRM activities include ongoing physical and virtual software patch and update management, high-availability and failover configuration, network zoning, routing, and related security tasks, as well as cabling and cable management.

IRM activities also entail creating and configuring virtual resources from physical resources—for example, establishing virtual servers and virtual machines, virtual storage (disk and tape based) and file systems, and virtual networks and I/O interfaces. Once virtual resources are created from configured physical resources, data and applications need to be moved and migrated to the available servers, storage, and network entities. For example, as part of routine IRM activities, migration or conversion of applications, servers, software, and data to virtual machiness involves converting guest operating systems and applications and moving them from physical to virtual servers. Another example is the movement of data and applications from an existing storage system to a new or upgraded storage solution. Other IRM activities involve physical-to-virtual, virtual-to-virtual, and virtual-to-physical resource movement and migration for routine maintenance, load-balancing, technology upgrades, or in support of business continuity and disaster recovery.

Data protection and security are important IRM tasks to ensure that data is available when needed but safely secured from various threat risks. Protecting data involves logical and physical security, including encryption and authentication as well as ensuring that copies exist, such as using snapshots, replication, and backup of data to meet service objectives. Another dimension of IRM activities across servers, storage, networks, and facilities is monitoring, managing, analyzing, and planning. These tasks involve resource usage monitoring, accounting, event notification, and reporting, as well as determining what resources can be consolidated and which ones need scaling to boost performance, capacity, or availability. Other tasks involve balancing to various service levels for different applications performance, availability, capacity, and energy, along with applicable reporting. Diagnostics, troubleshooting, event analysis, proactive resource management, and interdependency analysis between business functions and IT resources including asset and facilities management are also central IRM tasks.

4.2   Data Security (Logical and Physical)

There are many different threat risks for IT data centers, systems, applications, and the data they support. These range from acts of man to acts of nature and include technology failure, accidental, and intended threats. Many organizations feel that, other than malicious threats, their real threats are internal, and that how and who they hire and retain as employees is a differentiator. Most organizations agree that threats vary by application, business unit, and visibility. A common belief is that most threat risks are external; in reality, however, most threats (except for acts of nature) are internal. Some organizations believe that their firewalls and other barriers are strong enough to thwart attacks from outside. Some feel that firewalls and similar technology provide a false sense of security with little protection from internal threats.

Threats can be physical or logical, such as a data breach or virus. Different threat risks require multiple rings or layers of defenses for various applications, data, and IT resources, including physical security. Virtual data centers rely on both logical and physical security. Logical security includes access controls or user permissions for files, objects, documents, servers, and storage systems as well as authentication, authorization, and encryption of data. Another facet of logical security is the virtual or physical destruction of digital information known as digital shredding. For example, when a disk storage system, removable disk or tape cartridge, laptop or workstation is disposed of, digital shredding ensures that all recorded information has been securely removed. Logical security also includes how storage is allocated and mapped or masked to different servers, along with network security including zoning, routing, and firewalls.

Physical data protection includes securing facilities and equipment and access to management interfaces or workstations. Another dimension of physical security includes ensuring that data being moved or transported electronically over a network or physically is logically secured with encryption and physical safeguards including audit trails and tracking technology. For example, solutions are available today to retrofit existing magnetic tape and removable hard disk drives with external physical barcode labels that include embedded radio frequency identification (RFID) chips. The RFID chips can be used for rapid inventory of media being shipped to facilitate tracking and eliminate falsely reported lost media. Other enhancements include shipping canisters with global positioning system and other technologies to facilitate tracking during shipping.

In general, the overwhelming theme is that there is a perception that encryption key management is complex and this complexity is a barrier to implementation. Not protecting data, particularly data "in flight," with encryption due to fears of losing keys is similar to not locking your car or home because you might lose your keys. Key management solutions are available from different sources, with some solutions supporting multiple vendors' key formats and technologies.

Some organizations are exploring virtual desktop solutions as a means of moving away from potential desktop data exposure and vulnerabilities. Many organizations are racing to encrypt laptops as well as desktops. Some organizations limit USB ports to printer use only. Some organizations are also beefing up audit trails and logs to track what data was moved and copied where, when, and by whom. USB devices are seen as valuable tools, even given all of their risks, for moving and distributing data where networks don't exist or are not practical.

An evolving dimension to protecting data and securing virtual data centers is distributed remote offices along with traveling or telecommuting workers who occupy virtual offices. The threat risks can be the same as for a primary traditional data center and include others such as loss or theft of laptops, workstations, personal digital assistants (PDAs) or USB thumb drives containing sensitive information. When it comes to security, virtual data centers require multiple levels of logical and physical security across different technology domains.

4.3   Data Protection and Availability for Virtual Environments

If data is important enough to be backed-up or replicated, or if it needs to be archived for possible future use, then it is important enough to make multiple copies (including on different media types) at different locations.

There are many challenges but also options related to protecting data and applications in a virtual server environment. For example, in a non-virtualized server environment, the loss of a physical server will affect applications running on that server. In a highly aggregated or consolidated environment, the loss of a physical server supporting many virtual machines (VMs) will have a much more significant impact, affecting all the applications supported by the virtual servers. Consequently, in a virtual server environment, a sound data protection strategy is particularly important.

Popular approaches and technologies to implement server virtualization include Citrix/Xen, Microsoft, Virtual Iron, and VMware, as well as vendor-specific containers and partitions. Many data protection issues are consistent across different environments with specific terminology or nomenclature. Virtual server environments often provide tools to facilitate maintenance and basic data protection but lack tools for complete data protection, business continuity, or disaster recovery. Instead, virtual server vendors provide **application programming interfaces (APIs),** other tools, or solution/software development kits (SDKs) so that their partners can develop solutions for virtual and physical environments. For example, solutions from VMware and Virtual Iron include SDKs and APIs to support pre- and postprocessing actions for customization and integration with VMware Consolidated Backups (VCBs), VMotion, or LiveMigration from Virtual Iron.

4.3.1 Time to Re-Architect and Upgrade Data Protection

A good time to rethink data protection and archiving strategies of applications and systems data is when server consolidation is undertaken. Instead of simply moving the operating system and associated applications from a "tin"-wrapped physical server to a "software"-wrapped virtual server, consider how new techniques and technologies can be leveraged to improve performance, availability, and data protection. For example, an existing server with agent-based backup software installed sends data to a backup server over the LAN for data protection. However, when it is moved to a virtual server, the backup can be transitioned to a LAN-free and server-free backup. Thus LAN and other performance bottlenecks can be avoided.

From a historical data protection perspective, magnetic tape has been popular, cost effective, and the preferred data storage medium for retaining data. Recently, many organizations have been leveraging storage virtualization

Table 4.1   Data Protection Options for Virtual Server Environments

Capability	■ Characteristics	■ Description and Examples
Virtual machine (VM) migration	■ Move active or static VMs ■ Facilitate load balancing ■ Provides pro-active failover or movement as opposed to data recovery and protection	■ Vmotion, Xenmotion, LiveMigration ■ May be physical processor architecture dependent ■ Moves running VMs from server to server ■ Shared-access storage required along with another form of data protection to prevent data loss
Failover high availability (HA)	■ Proactive movement of VMs ■ Automatic failover for HA ■ Local or remote HA ■ Fault containment/isolation ■ Redundant Array of Independent Disks (RAID)-protected disk storage	■ Proactive move of a VM to a different server ■ Requires additional tools for data movement ■ Low-latency network bandwidth needed ■ Replication of VM and application-septic data ■ Isolate from device failure for data availability
Snapshots	■ Point-in-time (PIT) copies ■ Copies of current VM state ■ May be application aware	■ Facilitate rapid restart from crash or other incident ■ Guest operating system, VM, appliance, or storage system based ■ Combine with other forms of data protection

Table 4.1   Data Protection Options for Virtual Server Environments (continued)

Backup and restore	<ul><li>Application based</li><li>VM or guest operating system based</li><li>Console subsystem based</li><li>Proxy server based</li><li>Backup server or target resides as guest in a VM</li></ul>	<ul><li>Full image, incremental, differential, or file level</li><li>Operating system- and application-specific support</li><li>Agent or agent-less backup in different locations</li><li>Backup over LAN to server or backup device</li><li>Backup to local or SAN attached device</li><li>Proxy-based for LAN and server-free backup</li></ul>
Local and remote replication	<ul><li>Application based</li><li>VM or guest operating system based</li><li>Console subsystem based</li><li>External appliance based</li><li>Storage array based</li></ul>	<ul><li>Application or operating system based</li><li>Network, fabric, or storage system based</li><li>Application-aware snapshot integration</li><li>Synchronous for real-time low latency</li><li>Asynchronous for long distance, high latency</li></ul>
Archiving	<ul><li>Document management</li><li>Application based</li><li>File system based</li><li>Compliance or preservation</li></ul>	<ul><li>Structured (database), semistructured (email), and unstructured (files, PDFs, images, video)</li><li>Compliance or regulatory based</li><li>Noncompliance for long-term data retention</li></ul>

in the form of transparent access of disk-based backup and recovery solutions. These solutions emulate various tape devices and tape libraries, and coexist with installed backup software and procedures. Magnetic tape remains one of, if not the most, efficient data storage medium for inactive or archived data. Disk-to-disk (D2D) snapshots; backups, and replication have become popular options for near-term and real-time data protection.

With a continuing industry trend toward using D2D for more frequent and timely data protection, tape is finding a renewed role in larger, more infrequent backups for large-scale data protection in support of long-term archiving and data preservation. For example, D2D, combined with compression and de-duplication disk-based solutions, is used for local, daily, and recurring backups. Meanwhile, weekly or monthly full backups are sent to tape to free disk space as well as address PCFE concerns.

4.3.2 Technologies and Techniques—Virtual Server Data Protection Options

Just as there are many approaches and technologies to achieve server virtualization, there are many approaches for addressing data protection in a virtualized server environment. Table 4.1 provides an overview of data protection capabilities and characteristics in a virtualized server environment.

Complete and comprehensive data protection architectures should combine multiple techniques and technologies to meet various **recovery time objectives (RTOs)** and **recovery point objectives (RPO)**. For example, VM movement or migration tools such as VMware VMotion or Virtual Iron LiveMigration provide proactive movement for maintenance or other operational functions. These tools can be combined with third-party data movers, including replication solutions, to enable VM crash restart and recovery or basic availability. Such combinations assume that there are no issues with dissimilar physical hardware architectures in the virtualized environment.

Data protection factors to consider include:

- RTO and RPO requirements per application, VM/guest or physical server
- How much data changes per day, along with fine-grained application-aware data protection
- Performance and application service-level objectives per application and VM
- The distance over which the data and applications need to be protected
- The granularity of recovery needed (file, application, VM/guest, server, site)

■ Data retention as well as short-term and longer-term preservation (archive) needs

Another consideration when comparing data protection techniques, technologies, and implementations is application-aware data protection. Application aware data protection ensures that all data associated with an application, including software, configuration settings, data, and current state of the data or transactions, is preserved.

To achieve application-aware and comprehensive data protection, all data, including memory-resident buffers and caches pertaining to the current state of the application, needs to be written to disk. At a minimum, application-aware data protection involves quiescing (suspending) file systems and open files data to be written to disk prior to a snapshot, backup, or replication operation. Most VM environments provide tools and APIs to integrate with data protection tasks.

4.3.3   Virtual Machine Movement and Migration

Often mistaken, or perhaps even positioned, as data protection tools and facilities, virtual machine movement or migratory tools are targeted and designed for maintenance and proactive management. The primary focus of live VM migration tools is to be able to proactively move a running or active VM to a different physical server without disrupting service.

For example, VMotion can be used to maintain availability during planned server maintenance or upgrades or to shift workload to different servers based on expected activity or other events. The caveat with such migration facilities is that, while a running VM can be moved, the VM still needs to be able to access its virtual and physical data stores. This means that data files must also be relocated. It is important to consider how a VM movement or migration facility interacts with other data protection tools including snapshots, backup, and replication, as well as with other data movers.

In general, considerations in live movement facilities for VMs include:

■ How does a VM mover support dissimilar hardware architectures (e.g., Intel and AMD)?

- Is there a conversion tool (e.g., physical to virtual), or does it perform live movement?

- Can the migratory or movement tool work on both a local and wide area basis?

- How do tools interact with other data protection tools when data is moved with the VM?

- What are the ramifications of moving a VM and changes to Fibre Channel addressing?

- How many concurrent moves or migrations can take place at the same time?

- Is the movement limited to virtual file system-based VMs, or does it include raw devices?

4.3.4   High Availability

Virtual machine environments differ in their specific supported features for **high availability (HA),** ranging from the ability to failover or restart a VM on a different physical server to the ability to move a running VM from one physical server to another physical server (as discussed in the previous section). Other elements of HA for physical and virtual environments include eliminating single points of failure to isolate and contain faults, for example, by using multiple network adapters, redundant storage I/O host bus adapters, and clustered servers.

A common approach for HA data accessibility is RAID-enabled disk storage to protect against data loss in the event of a disk drive failure. For added data protection, RAID data protection can be complemented with local and remote data mirroring or replication to protect against loss of data access due to a device, storage system, or disk drive failure. RAID and mirroring, however, are not a substitute for backup, snapshots, or other point-in-time discrete copy operations that establish a recovery point.

RAID provides protection in the event of disk drive failures; RAID does not by itself protect data in the event that an entire storage system is damaged. While replication and mirroring can protect data if a storage system is destroyed or lost at one location, if data is deleted or corrupted at one location, that action will be replicated or mirrored to the alternate copy. Consequently, some form of time interval-based data protection, such as a snapshot or backup, needs to be combined with RAID and replication for a comprehensive and complete data protection solution.

4.3.5 Snapshots

There are a number of reasons why snapshots, also known as **point-in-time (PIT)** copies and associated technologies might be utilized. Snapshots create a virtual backup window to enable data protection when a physical backup window is shrinking or no longer exists. Snapshots provide a way of creating virtual time to get essential data protection completed while minimizing impacts to applications and boosting productivity. Different applications have varying data protection requirements, including RTO, RPO, and data retention needs. Other reasons for making snapshots include making copies of data for test purposes, including software development, regression testing, and disaster recovery testing; making copies of data for application processing, including data warehouse, data marts, reporting, and data mining; and making copies to facilitate non-disruptive backups and data migration.

Snapshots can reduce downtime or disruptions associated with traditional data protection approaches such as backup. Snapshots vary in their implementation and location, with some being full copies while others are "delta-based." For example, an initial full copy is made with deltas or changes recorded, similar to a transaction or redo log, with each snapshot being a new delta or point-in-time view of the data being protected. Snapshot implementations can also vary in where and how the snapshot data is stored.

Because snapshots can take place very quickly, an application, operating system, or VM can be quiecesed, a quick snapshot taken of the current state at that point in time, and then resume with normal processing. Snapshots work well for reducing downtime as well as speeding up backups. Snapshots reduce the performance impact of traditional backups by only copying changed data, similar to an incremental or differential backup but on a much more granular basis. Snapshots can be made available to other servers in a shared storage environment to further off-load data protection. An example is using a proxy or backup server to mount and read the snapshots to construct an offline backup.

For virtual environments, snapshots can be taken at the VM or operating system layer, with specific features and functionalities varying by vendor implementation. Snapshots can also be taken in storage systems that are integrated with a guest operating system, applications, or VM. Snapshots can also be taken in network- or fabric-based appliances that intercept I/O data streams between servers and storage devices.

One of the key points in utilizing snapshots is to make sure that when a snapshot is taken, the data that is captured is the data that was expected to be recorded. For example, if data is still in memory or buffers, that data may not be flushed to disk files and captured. Thus, with fine-grained snapshots, also known as near or coarse continuous data protection (CDP), as well as with real-time fine-grained CDP and replication, 100% of the data on disk may be captured. But if a key piece of information is still in memory and not yet written to disk, critical data to ensure and maintain application state coherency and transaction integrity is not preserved. While snapshots enable rapid backup of data as of a point in time, snapshots do not by themselves provide protection in the event of a storage system failure; thus, snapshots need to be backed up to another device.

4.3.6   Agent-Based and Agent-Less Data Protection

Agent-based backup, also known as LAN-based backup, is a common means of backing up physical servers over a LAN. The term comes from the fact that a backup agent (backup software) is installed on a server, with the backup data being sent over a LAN to a backup server or to a locally attached tape or disk backup device.

Given the familiarity with and established existing procedures for using LAN- and agent-based backup, a first step for data protection in a virtual server environment may be to simply leverage agent-based backup while re-architecting virtual server data protection.

Agent-based backups as shown in Figure 4.2 are relatively easy to deploy, as they may be in use for backing up the servers being migrated to a virtual environment. Their main drawback is that they consume physical memory, CPU, and I/O resources, causing contention for LAN traffic and impacting other VMs and guests on the same virtualized server.

Backup client or agent software can also have extensions to support specific applications such as Exchange, Oracle, SQL, or other structured data applications as well as being able to handle open files or synchronize with snapshots. One of the considerations in using agent-based backups, however, is what support exists for the backup devices or targets. For example, are locally attached devices supported from an agent, and how can data be moved to a backup server over a network in a LAN-friendly and efficient manner?

Physical servers, when running backups, have to stay within prescribed backup windows while avoiding performance contention with

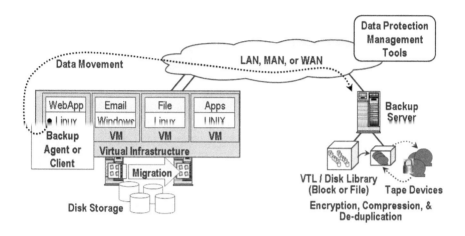

Figure 4.2   Agent-Based Backup over a LAN

other applications and also avoiding LAN traffic contention. In a consolidated virtual server environment, multiple competing backup jobs may also vie for the same backup window and server resources. Care needs to be exercised when consolidating servers into a virtual environment to avoid performance conflicts and bottlenecks.

4.3.7   Proxy-Based Backup

Agent- or client-based backups running on guest operating systems consume physical resources, including CPU, memory, and I/O, resulting in performance challenges for the server and the LAN (assuming a LAN backup). Similarly, an agent-based backup to a locally attached disk, tape, or virtual tape library (VTL) will still consume server resources, resulting in performance contention with other VMs or other concurrently running backups.

In a regular backup, the client or agent backup software, when requested, reads data to be backed up and transmits the data to the target backup server or storage device while also performing associated management and record keeping tasks. Similarly, during restore operations, the backup client or agent software works with the backup server to retrieve data based on the specific request. Consequently, the backup operation places a demand burden on the physical processor (CPU) of the server while consuming memory and I/O bandwidth. These competing demands can and need to be managed if multiple backups are running on the same guest operating system and VM or on different VMs.

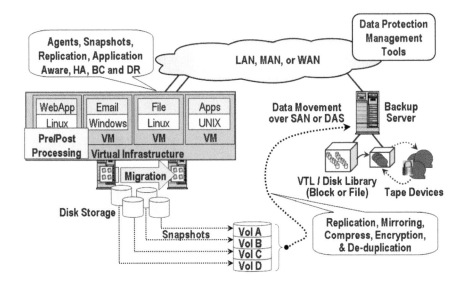

Figure 4.3   VMware VCB Proxy-Based Backup Example

One approach to addressing consolidated backup contention is to leverage a backup server and configure it as a proxy (see Figure 4.3) to perform the data movement and backup functions. Proxy backups work by integrating with snapshot, application, and guest operating system tools for pre- and postprocessing. As an example, VMware Consolidated Backup (VCB) is a set of tools and interfaces that enable a VM, its guest operating system, applications, and data to be backed up by a proxy while reducing the CPU, memory, and I/O resource consumption of the physical server compared to a traditional backup.

VCB is not a backup package. Rather, it is an interface to VMware tools and enables third-party backup and data protection products to work. To provide data protection using VCB, third-party backup tools are required to provide scheduling, media, and backup management. Third-party tools also manage the creation of data copies or redirect data to other storage devices, such as virtual tape libraries and disk libraries, equipped with compression and data de-duplication to reduce data footprint. VM virtual disk images are sparse or hollow, meaning that there is a large amount of empty or blank space with many similar files that lend themselves to being compressed and de-duplicated.

In addition to off-loading the physical server during the proxy backup, LAN traffic is not affected, as data can be moved or accessed via a

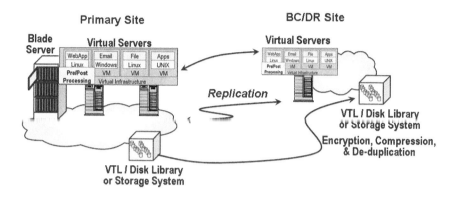

Figure 4.4   Data Replication for High Availability and Data Protection

shared storage interconnect. Third-party backup and data protection software on a proxy server can also perform other tasks, including replicating the data to another location, keeping a local copy of the backup on disk-based media with a copy at the remote site on disk, as well as on a remote offline tape if needed.

4.3.8   Local and Remote Data Replication

There are many approaches to data replication and mirroring, shown generically in Figure 4.4, for local and remote implementations to address different needs, requirements, and preferences.

An important consideration in data mirroring and replication is distance and latency, which may result in data delay and negative performance impacts. Distance is a concern, of course, but the real enemy of synchronous data movement and real-time data replication without performance compromise is latency. There is a common perception that distance is the enemy of synchronous data movement. Generally speaking, latency increases over distance; thus, the common thinking is that distance is the problem in synchronous data movement. The reality is that even over relatively short distances, latency can negatively impact synchronous real-time data replication and data movement.

Distance and latency bear on replication and data movement by affecting decisions as to whether to use synchronous or asynchronous data movement methods. Besides cost, the major factors are performance and data protection. Synchronous data transfer methods facilitate real-time data protection, enabling a recovery point objective (RPO) of near zero. However, the trade-off is that over distance, or, high-latency networks, application performance is negatively impacted while the system waits for remote I/O operations to be completed.

Another approach to the problem is to use asynchronous data transfer modes, in which a time delay is introduced along with buffering. By using a time delay and buffering, application performance is not affected, as I/O operations appear to applications as having completed. The trade-off with asynchronous data transfer modes is that although performance is not degraded over long-distance or high-latency networks, there is a larger RPO exposure potential for data loss while data is in buffers waiting to be written to remote sites.

A combination of synchronous and asynchronous data transfer may be used, providing a tiered data protection approach—for example, using synchronous data transfer for time-critical data to a reasonably nearby facility over a low-latency network, and replicating less critical data asynchronously to a primary or alternative location farther away. A hybrid approach is to perform synchronous data replication to a nearby facility, then perform a second, asynchronous replication to another site farther away.

A general caveat is that replication by itself does not provide complete data protection; replication is primarily for data availability and accessibility in the event of a component, device, system, or site loss. Replication should be combined with snapshots and other point-in-time discrete backup data protection to ensure that data can be recovered or restored to a specific RPO. For example, if data is corrupted or deleted on a primary storage device, replication will replicate the corruption or deletion to alternate sites, hence the importance of being able to recover to specific time intervals for rollback.

4.3.9   Archiving and Data Preservation

Data preservation or archiving of structured (database), semistructured (email and attachments), and unstructured (file-oriented) data is an effective means to reduce data footprint and associated PCFE, backup/recovery, business continuity, disaster recovery, and compliance issues. Given the current focus on addressing PCFE-associated issues and the growing awareness of the need to preserve data offline or near-line to meet regulatory requirements, magnetic tape is an effective complementary technology to D2D backups. Magnetic tape continues to be a strong solution for its long-term cost and performance and its effective offline data preservation.

4.3.10   Complete Data Protection

Figure 4.5 shows a combination of data protection techniques including a storage area network (SAN) or network attached storage (NAS) system with RAID for data availability, local and remote data replication, snapshots to facilitate high-speed data backups, D2D, and tape-based backups with encryption. In addition, compression and de-duplication can be incorporated to help reduce data footprint and the physical storage space required, or to store more data in the same space.

Virtual backups, that is, backups to a remote managed service provider (MSP), software or storage as a service (SaaS), or cloud-based service can also be part of a complete data protection approach. For example, remote office and branch offices (ROBO) or home-based and traveling workers (virtual offices) can be backed up, replicated, and have their data protected to either an internal managed service (e.g., a virtual data center) or to a third-party service. For smaller organizations, third-party backup and data protection MSPs and cloud-based services or SaaS-based solutions are an effective means for enhancing timely and affordable data protection.

Things to consider when evaluating data protection techniques include:

- Potential threats to applications and data
- RTO and RPO requirements for applications and associated data
- Who will perform data recovery
- How transparent the data protection and recovery scheme need to be
- Technologies currently in place (hardware, software, etc.)
- Alternative techniques and technologies for data protection
- Budget, timeframe, tolerance to disruption, and risk aversion
- Which solutions are best for different applications
- Availability of experienced help for assessment, validation, or implementation

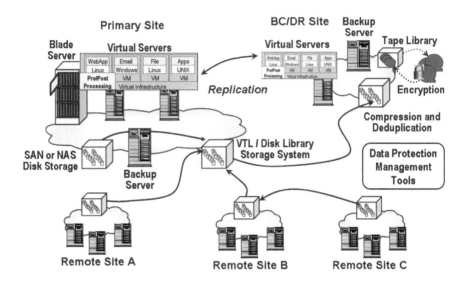

Figure 4.5   Local and Remote Data Protection

4.4   Data Protection Management and Event Correlation

Data protection management (DPM) has evolved from first-generation backup reporting technology to incorporate multivendor and cross-technology domain capabilities. In addition, DPM tools are evolving to manage multiple aspects of data protection along with event correlation.

Some DPM tools are essentially reporting, status, or event monitoring facilities that provide passive insight into what is happening in one or more areas of infrastructure resource management (IRM). Other DPM tools provide passive reporting along with active analysis and event correlation, providing a level of automation for larger environments. Cross-technology domain event correlation connects reports from various IT resources to transform fragments of event activity into useful information on how, where, why, and by whom resources are being used. In virtualized environments, given the many different interdependencies, cross-technology domain event correlation is even more valuable for looking at end-to-end IRM activities

Increasing regulatory requirements combined with pressure to meet service levels and 24-7 data availability has resulted in data protection interdependencies across different business, application, and IT entities.

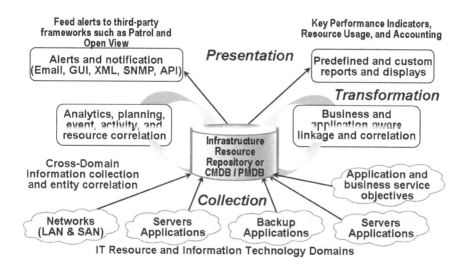

Figure 4.6   Data Protection Management Across Technology Domains

Consequently, timely and effective DPM requires business and application awareness to correlate and analyze events that affect service and IT resource usage. Business awareness is the ability to collect and correlate IT assets to application interdependencies and resource usage with specific business owners or functions for reporting and analysis. Application awareness is the ability to relate IT resources to specific applications within the data protection environment to enable analysis and reporting.

A challenge in business- and application-aware DPM has been the fact that many organizations maintain information about business units, applications, IT resources, or asset ownership and usage in disparate database and repository formats. For example, information is kept in configuration management databases (CMDB), performance management databases (PMDB), and metadata repositories, among other locations. To support business- and application-aware data protection, DPM tools need to support access of external sources.

Effective DPM includes knowing what, where, when, why, and by whom resources are being used to deliver service and how effectively that service is being delivered (Figure 4.6). To enable timely IRM across different technology domains and disciplines, including DPM, automated data collection and correlation of event and activity information is needed. Event correlation from different sources facilitates root-cause analysis so that service levels and compliance objectives can be meet. With a focus on reducing electrical energy consumption and associated environmental impacts, DPM can be used to ensure that data protection resources are being used optimally.

Although an environment may have multiple tools and technologies to support IRM activities, DPM tools are evolving to support or coexist with management of multiple data protection techniques, including backup (to disk or tape), local and remote mirroring or replication, snapshots, and file systems. Key to supporting multiple data protection approaches and technologies is the ability to scale and process in a timely manner large amounts of event and activity log information. At the heart of a new breed of IRM tools, including DPM solutions, are robust cross-technology resource analysis and correlation engines to sift disparate data protection activity and event logs for interrelated information.

Examples of products that have event correlation as their basis with different personalities (capacity management, compliance coverage, data protection management, configuration validation, replication management, network management) include Akorri, Continuity, EMC (Smarts), Onaro (now part of NetApp), and WysDM (now part of EMC). While the mentioned tools perform a specific or multiple IRM-related functions, they all support cross-technology domain event correlation.

4.5 Server, Storage, and Network Resource Management

Performance and capacity planning can be combined as complementary activities along with server or **storage resource management (SRM)** and utilization, or they may be handled as separate tasks. Performance tuning and optimization may initially be seen as reactionary tasks to respond to specific situations. A performance plan and ongoing performance tuning initiative can support a shift from reactionary to tactical and longer-term strategic management approaches. For example, shifting to a performance plan approach, in which performance and usage are analyzed and optimized as part of an overall growth plan, can help maximize and optimize spending.

IRM reporting and monitoring tools should allow an IT administrator to see across different technology domains and from virtual server to physical storage for the full IRM picture. In addition, capacity and resource usage tools add performance or activity reporting to traditional space or capacity utilization to provide a more holistic view of resource usage. Performance and utilization should be evaluated in tandem. It's bad policy to scale up utilization only to find that performance is suffering.

4.5.1   Search and eDiscovery

Data classification and search tools have several functions, including discovery, classification, and indexing, as well as searching, reporting, and taking action on discovered data—for example, identifying what files to migrate from active online storage to offline storage for archive purposes. For compliance-related data, taking action includes marking data for litigation hold to prevent tampering or taking action such as deleting data based on policies. In general, taking action refers to the ability to interface with various storage systems (online, near-line, and offline), including object-based and archiving systems, to enable management and migration of data.

Storage resource management and basic data discovery and classification tools include file path and file meta data discovery tools. SRM tools have a vertical focus on storage and file identification for storage management purposes, including allocation, performance, and reporting. Some tools provide basic SRM-like functionality along with more advanced capabilities including archiving, document management, email, and data migration capabilities. Deep content discovery, indexing, classification, and analysis tools support features such as word relativity, advanced language support, and search and discovery features for vertical markets.

When looking at data discovery and indexing tools, the intended and primary use of the technology should be kept in mind. For example, is the planned use of the tools to perform deep content discovery for compliance, legal litigation, and intellectual property search? Perhaps you are looking to identify what files exist, when they were last accessed, and what might be candidates for moving to different tiers of storage. By keeping primary objectives in focus, you may find that different tools work better for various tasks, and that more than one tool is needed.

Architectural considerations include performance, capacity, and depth of coverage, along with discovery, security and audit trails. Policy management should be considered, along with policy execution, interfaces with other policy mangers, and data migration tools. Some tools also support interfaces to different storage systems such as vendor-specific APIs for archiving and compliance storage. Consider whether the candidate tools have embedded or built-in support for processing different templates, lexicons, syntax, and taxonomies associated with different industries and regulations. For example, when dealing with financial documents, the tool should support processing of data in the context of various financial taxonomies such as banking, trading, benefits, and insurance, among others. If legal documents are being processed, then support for legal taxonomies will be needed.

Classifying data is complex, and for some services providers who merely "house" data, the actual value of the data may not be known. Although tools exist, they are limited in their extensiveness and scalability. Interaction with lines of business and those developing the applications are important to understand the value of data. Tiered security is needed, but a methodology also needs to exist and be tied to data value, location, and line of business.

Understanding target applications and needs for discovery tools will help to ensure a positive and successful solution. To understand what files exist on a system to help implement a tiered storage environment, start by looking at traditional SRM-type tools. If, on the other hand, deep data discovery is needed to support litigation, compliance, or other functions, then consider more advanced tools. Some tools can meet multiple objectives, but it is important to understand what other aspects of a system may be affected.

4.5.2   Rescuing Stranded or Orphaned Resources

Orphaned storage is any data, file, table space, object, file system, LUN, physical volume, or storage device that appears to be in use but has been abandoned or forgotten. Orphaned storage can result from application or system errors that have not been cleared after a restart, system maintenance, upgrade, or other activity. Orphaned storage can exist in different forms and in various locations in most environments, ranging from orphaned data in a database to orphaned storage in an email system or orphaned files in networked attached storage (NAS) or traditional file system. Orphaned storage can also exist in the form of unused or unallocated LUNs, physical volumes, or even individual disk drives.

One challenge in finding orphaned storage and data is determining whether data is, in fact, orphaned. Some files may appear to be orphaned, but they may be in line to be archived, in which case they should be migrated to some other storage medium. Likewise, some data files or storage volumes may appear to be allocated, having not been released after some previous use. The data or storage may not have been de-allocated if someone forgot to tell someone else that the storage is no longer being used, or some documentation somewhere was not updated to indicate that the storage can be de-allocated and re-provisioned. Another cause of orphaned storage is system or application error. For example, over a period of time, inconsistencies can appear in databases or file systems that require a repair operation to free up unused, yet allocated, storage and index pointers.

Consider the following for finding and eliminating or adopting orphaned storage:

- Clean up temporary, scratch, and work space on a regular basis.
- Run database, application-specific, and file system consistency checks.
- Utilize vendor tools or have the vendor check for orphaned devices.
- Leverage discovery and SRM tools to verify how storage is being used.
- Use configuration analysis tools to validate storage configurations.
- Look for files that appeared around the time a system or application error occurred.
- Have database administrators (DBAa) check for duplicate data, orphaned rows or tables.
- Apply policies as part of a clean-up after upgrades to find orphaned storage.

4.5.3   Capacity, Availability, and Performance Planning

There may not appear to be a link between availability and performance and capacity planning, but there is a direct connection. If a resource is not available, performance is affected. And if a resource has poor performance or limited supply, availability and accessibility are affected.

Capacity planning and capacity management are used in a variety of businesses. In a manufacturing company, for example, they are used to manage inventory and raw goods. Airlines use capacity planning and capacity management to determine when to buy more aircraft. Electric companies use them to decide when to build power plants and transmission networks. In the same way, IT departments use capacity planning and capacity management to derive maximum value and use from servers, storage, networks, and facilities while meeting service-level objectives or requirements.

Consider some common questions and comments with regard to performance and capacity planning:

- Hardware is cheap; why tie someone up doing capacity planning and tuning?

  - While hardware is becoming less expensive, power, cooling, floor space, and environmental (PCFE) resources are increasing in cost and decreasing in availability.

- People are already busy if not overworked; why give them more to do?

  - With planning, resources (people, hardware, software, networks, and budget) can be utilized more effectively to address and maximize available PCFE resources.

- Why not buy more hardware and have a vendor manage it?

  - This may be an alternative if it is viable from cost and business perspectives. However, adding more resources adds to the expense to manage, protect, and utilize the resources as well as space to house the resources and power to operate and cool them.

Capacity and performance planning should address peak processing periods and degraded performance whether planned, scheduled, or unexpected. Performance and capacity planning activities have occurred in the enterprise environments of S/390 mainframes and open systems platforms for many years. Historically, capacity planning activities have for the most part focused on large (expensive) components, including processors, memory, network interconnects, and storage subsystems (disk and tape) for larger enterprise environments.

Capacity planning can be a one-time exercise to determine how much and what types of resources are needed to support a given application. A nontactical approach to resource needs assessment and sizing is simply to acquire some amount of resources (hardware, software, networks, and people) and buy more as needed. A strategic approach might evolve from the tactical to make more informed decisions and timed acquisitions. For example, knowing your resource needs ahead of time, you might be able to take advantage of special vendor incentives to acquire equipment that suits your needs on your terms. Similarly, if the terms are not favorable and resource usage is following the plan, you may choose to delay your purchase.

Virtual data centers help to abstract physical resources from applications and users. However, increased complexity needs to be offset with end-to-end diagnostics and assessment tools along with proactive event correlation and analysis tools. Having adequate resources when needed to sustain business growth and meet application service requirements is a balancing act. The balance is having enough server, storage, networking, and PCFE resources on hand without having too much, resulting in higher costs, or not enough, resulting in poor service.

Poor metrics and insight can lead to poor decisions and management. Look at servers from more than a percent utilization viewpoint, considering also response time and availability. Look at storage from an IOPS and bandwidth performance perspective along with response time or latency as well as available capacity. Look at networking from a latency standpoint in addition to cost per given bandwidth and percent utilization.

If you are new to capacity planning, check out the Computer Measurement Group (CMG), which is focused on cross-technology, vendor- and platform-neutral performance, and capacity planning management. In general, the recommendation is to start simple and build on existing or available experience and skill. Identify opportunities that will maximize positive results to gain buy-in and evolve to more advanced scenarios.

4.5.4   Energy Efficiency and PCFE Management Software

Power management and reporting software tools and appliances help with reporting of energy usage by various IT resources. Some solutions are able to interact and take proactive or reactive action to manage energy usage on a server, storage, network, or application basis. Solutions vary in robustness, with some able to apply business rules and polices on an application basis or technology basis (e.g. server, storage, or network).

For example, a power management solution based on set polices, including a weekly or monthly schedule, can instruct servers to go into a low-power mode or to interact with server virtualization software to shift virtual machines and applications to different servers to meet PCFE objectives. Similar to the electrical power generation and transmission environment discussed in Part I, which relies on command and control software and tools, active power management and intelligent power management tools continue to evolve to enhance command and control for energy optimization in IT data center environments.

4.6 Summary

There are many vendors with solutions to address various aspects of infrastructure resource management in a physical or virtual data center. Examples include BMC, Brocade, Cisco, Egenera, EMC, Emerson, HP, IBM, LSI, Microsoft, NetApp, Novell, Opalis, Racemi, Scalent, Sun, Symantec, Teamquest, Tek-Tools, Uptime, Vizoncore, and VMware, among others.

The benefits of server virtualization for consolidation as well as management transparency are becoming well understood, as are the issues associated with protecting data in virtualized server environments. There are many options to meet different recovery time objective and recovery point objective requirements. Virtualized server environments or infrastructures have varying functionalities and interfaces for application-aware integration to enable complete and comprehensive data protection with data and transactional integrity.

A combination of tape- and disk-based data protection, including archiving for data preservation, coupled with a data footprint reduction strategy, can help to address power, cooling, floor space, and environmental needs and issues. There is no time like the present to reassess, re-architect, and reconfigure your data protection environment, particularly if you are planning on or have already initiated a server virtualization initiative. Virtual server environments require real and physical data protection. After all, you cannot go forward from a disaster or loss of data if you cannot go back to a particular point in time and recover, restore, and restart, regardless of your business or organization size.

Chapter 5

Measurement, Metrics, and Management of IT Resources

You can't manage what you don't know about or have insight into.

In this chapter you will learn:

- Why metrics and measurements are important for understanding how resources are used
- How to ensure that required resources are available
- Key tenets of metrics and benchmarks for different use scenarios

Good decision making requires timely and insightful information. Key to making informed decisions is having insight into IT resource usage and services being delivered. Information about what IT resources exist, how they are being used, and the level of service being delivered is essential to identifying areas of improvement to boost productivity and efficiency and to reduce costs. This chapter identifies metrics and techniques that enable timely and informed decisions to boost efficiency and meet IT service requirements.

Metrics reflect IT resource usage across different dimensions, including performance, availability, capacity, and energy, for active and inactive periods over different time frames. IT resources can be described as being either active or performing useful work or inactive when no work is being performed. In keeping with the idea that IT data centers are information factories, metrics and measurements are similar to those for other factories. Factories in general, and highly automated ones in particular, involve costly resources and raw goods being used to create valuable goods or services, all of which need to be measured and tracked for effective management.

The overall health and status of the equipment, steady supply of raw materials, energy or power supply, and quality of service are constantly measured for timely management decisions to be made. For example, in an electrical power plant, control rooms or operations nerve centers closely monitor different aspects of the facility and its resources, including fuel supply and cost, production schedules or forecasts, boilers, exhaust air scrubbers, water chillers and associated pumps, turbines and transmission status.

At a higher level, the generating power plant is being monitored by a network operations control center as part of the larger power transmission grid, similar to an IT data center being monitored by a network and operations control center. Both real-time and historical baseline information are used to manage generation supply proactively in order to meet current and predicted demand as well as to aid in future planning and maintenance purposes and to support customer usage and billing systems.

IT data centers are measured to gauge the health, status, efficiency, and productivity of resource usage to delivery a given level of service. Several types of metrics and measurements, ranging from component level to facility-wide, serve different purposes and audiences at various times. Metrics and measurements feed management tools, graphical user interfaces, dashboard displays, email and other notification mechanisms, frameworks and other monitoring tools with key performance indicators. Metrics and measurements provide insight about what is occurring, how resources are being used and the efficiency of that usage, and the quality of service. Metrics and measurements of resources include performance, availability, capacity and energy for servers, storage, networks, and facilities to meet a given level of service and cost objectives. Measurements can be used for real-time reactive and proactive monitoring and management of resources, event correlation for problem diagnosis and resolution, and for planning and analysis purposes, as shown in Figure 5.1.

There are different points of interest for different audiences at varying times during a product or technology lifecycle, as shown in Figure 5.1. For example, a vendor's engineers use comparative or diagnostic measurements at the component and system levels during research and development and during manufacturing and quality assurance testing. Performance and availability benchmarks, along with environmental power, cooling, and other metrics, are used for comparison and competitive positioning during the sales cycle. Metrics are also used on an ongoing basis to assess the health and status of how a technology is performing to meet expectations.

In the typical data center, a different set of metrics is used than what a vendor utilizes during design and manufacture. For example, instead of a

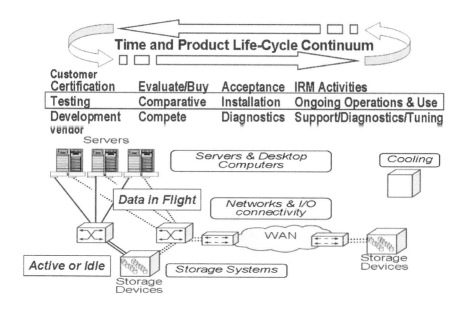

Figure 5.1   Metrics and Measurements Points of Interest—The Big Picture

detailed component focus, data center personnel generally take a more holistic view. That is, for a storage system, they may focus on the total amount of power being used, performance in terms of input/output operations per second (IOPS) or bandwidth, and available capacity in a given footprint.

Additional detailed information is generally available, such as how a storage system is performing at the disk drive or other component level. Similarly, vendors often have additional built-in measurement and diagnostic capabilities—sometimes hidden from customer access—that provide additional insight into activities and events for performance or troubleshooting purposes. Another example from a facilities perspective is the total power being used for the data center measured independently of work being performed or data being stored or even the type and quantity of IT resources such as servers, storage, and networking devices.

Electrical power usage will vary depending on the type of device, its design, and low-power features such as intelligent power management, adaptive voltage scaling, or adaptive power management. A device may require more power at start-up or power-up than when running in a steady state. Less power is generally needed when doing minimum or no work, and even less power when in a low-power (idle, standby, or sleep) mode.

A common example is a workstation or laptop computer using power management; the amount of power can be varied depending on activity. When active, power management features in the workstation operating system software can speed up or slow down the process to vary energy consumption. Other steps that can be taken on a workstation or laptop computer include dimming or turning off monitors, turning off or reducing power on disk drives, or entering a standby sleep mode or going into hibernation in addition to being completely powered off. Note that on a laptop computer, standby mode enables fairly rapid resumption of work or activity but draws less power running in normal mode.

Standby mode draws more power than when the computer is in hibernation mode, which, in turn, draws more power than if the device is completely powered off. In hibernation the device appears to be drawing no power; however, a very small amount of power is being used. The energy efficiency benefits of putting a laptop computer into standby or hibernation mode along with leveraging other power management features are most commonly used to maximize the amount of time work that can be done when running on battery power.

A server is basically either busy doing work or is idle. Likewise, a networking device is generally either supporting movement of data between users and servers, between servers and other servers, between servers and storage, or between storage devices on a local, metropolitan, or wide-area basis. Storage devices support active work including movement of data to satisfy I/O operations such as read and write requests as well as storing data. Consequently, storage devices can be measured on an active or working basis as well as on the ability to store information over long periods of time.

Not all storage devices should be compared on the same active-or-idle workload basis. For example, magnetic tape devices typically store inactive data offline and consume power only when data is being read or written. By comparison, magnetic hard disk drives (HDD) are generally always spinning, ready or already performing work in data center storage systems. More information about different types and tiers of storage is given in Chapter 8.

5.1   Data Center-Related Metrics

There are many different sources of metrics and measurements in a typical IT data center. Metrics can be focused on activity and productivity, usually reported in terms of performance and availability, along with resource usage and efficiency. Consequently, metrics can reflect active work being done to deliver IT services in a given response time or work rate as well as indicate how IT resources are being used. IT equipment can be active or inactive. Data storage can be considered active when responding to read and write requires or I/O operations, and how much data is being stored can be measured independent of activity. Consequently, different types of storage need to be compared on different bases.

For example, magnetic tape used for of-line archival and backup data can be measured on an idle or inactive basis, such as capacity (raw, usable, or effective) per watt in a given footprint, data protection level, and price. While capacity is a concern, online active storage for primary and secondary data, including bulk storage, needs to be compared on an activity-per-watt basis along with how much capacity is available at a given data protection level and cost point to meet service-level objectives.

Metrics are available from a variety of sources in a data center. Different metrics have meaning to various groups, and multiple similar metrics may be collected by server, storage, and networking groups as well as separate metrics by facilities personnel.

Some common key attributes of all these measurements are that they are:

- Applicable to the function or focus area of various persons using the metrics
- Relevant to a given point or period in time for real-time or historical purposes
- Reproducible over different scenarios and relevant time periods
- Reflective of the work, activity, or function being performed

For servers, useful metrics include application response time, I/O queues, number of transactions, web pages or email messages processed, or other activity indicators as well as utilization of CPUs, memory, I/O or networking interfaces, and any local disk storage. Availability can be measured in terms of planned and unplanned outages or for different time frames such as prime time, nights, and weekends, or on a seasonal basis.

A common focus, particularly for environments looking to use virtualization for server consolidation, is server utilization. Server utilization does provide a partial picture; however, it is important to look also at performance and availability for additional insight into how a server is running. For example, a server may operate at a given low utilization rate to meet application service-level response time or performance requirements. For networks, including switches, routers, bridges, gateways, and other specialized appliances, several metrics may be considered, including usage or utilization; performance in terms of number of frames, packets, IOPS, or bandwidth per second; and latency, errors, or queues indicating network congestion or bottlenecks.

From a storage standpoint, metrics should reflect performance in terms of IOPS, bandwidth, and latency for various types of workloads. Availability metrics reflect how much time, or what percent of time, the storage is available or ready for use. Capacity metrics reflect how much or what percent of a storage system is being used. Energy metrics can be combined with performance, availability, and capacity metrics to determine energy efficiency. Storage system capacity metrics should also reflect various native storage capacities in terms of raw, unconfigured, and configured capacity. Storage granularity can be assessed on a total usable storage system (block, file, and object/cas) disk or tape basis or on a media enclosure basis—for example, disk shelve enclosure or individual device (spindle) basis. Another dimension is the footprint of the storage solution, such as the floor space and rack space and may include height, weight, width, depth, or number of floor tiles.

Measuring IT resources across different types of resources, including multiple tiers, categories, types, functions, and cost (price bands) of servers, storage, and networking technologies, is not a trivial task. However, IT resource metrics can be addressed over time to address performance, availability, capacity, and energy to achieve a given level of work or service delivered under different conditions.

It is important to avoid trying to do too much with a single or limited metric that compares too many different facets of resource usage. For example, simply comparing all IT equipment from an inactive, idle perspective does not reflect productivity and energy efficiency for doing useful work. Likewise, not considering low-power modes ignores energy-saving opportunities during low-activity periods. Focusing only on storage or server utilization or capacity per given footprint does not tell how much useful work can be done in that footprint per unit of energy at a given cost and service delivery level.

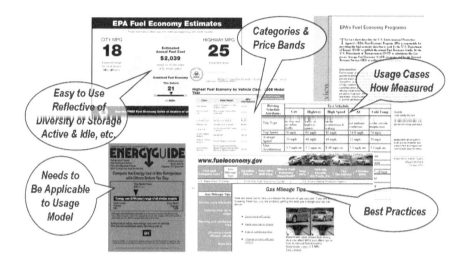

Figure 5.2   Metrics and Measurements for Different Categories, Uses, and Activities

Figure 5.2 shows an example of what can be done with metrics and measurements and what continues to be refined. Why should one compare IT resources to automobiles? Because both consume energy to do some amount of useful work, some carry cargo or passengers over long distances while others do frequent start-and-stop operations, some are always in use while others sit at rest consuming no power for extended periods of time. Both involve different types and categories of vehicles to address different tasks or functions.

You are probably familiar with comparisons of automobile miles per gallon for city and highway driving, but the U.S. Environmental Protection Agency (EPA) also provides breakdowns for different categories, price bands, and tiers of vehicles, from trucks and busses to automobiles. In addition, these vehicles are compared under various scenarios, such as city start-and-stop versus highway driving, highway driving at high speed, and highway driving with the air conditioning turned on.

While not 100% applicable, the comparison of different vehicles and useful work to energy consumed does have an interesting resemblance to similar comparisons of IT resources and is certainly food for thought. In fact, the EPA is currently working with the IT industry (vendors, customers, analysts, and consultants) to extend its Energy Star™ program, which already addresses consumer appliances and some consumer electronics including workstation and laptops. By the time you read this, the EPA should have in place the first specifications for Energy Star™ servers and be on its way to defining follow-on server specifications as well as specifications for Energy Star™ storage and data centers. (Note that while the fuel economy estimates and Energy Star are both EPA programs, they are overseen by different division within the EPA.)

The EPA is not alone in working on energy efficiency metrics. Other countries around the world are also looking into new metrics to support emission tax schemes and other initiatives.

5.2   Different Metrics for Different Audiences

What metrics and measurements are needed depend on the audience that will use them. For example, manufacturers or software developers have different needs and focus areas than sales and marketing departments. IT organizations may need data on the macro or "big picture" view of costs, energy usage, number or type of servers, amount of storage capacity, or how many transactions can be supported, among others. Needed metrics also vary at different times, including during development, manufacture and quality assurance, sales and marketing, customer acquisition and installation, integration testing, and ongoing support.

Metrics encompass performance, availability, capacity, and energy consumption along with general health and status, including the effectiveness of a solution. Given various usage scenarios, metrics address IT resources when being used to process work or move data as well as how data is stored when not in use. From the standpoint of power, cooling, floor space, and environment (PCFE), IT resources need to be monitored in terms of effective work along as well as how much data can be stored in a given footprint for a given level of service and cost.

Figure 5.3 shows on the left various points of focus for metrics, ranging from facilities at the bottom to business and application centered at the top. Various types of metrics and measurements are available in the different categories that feed to infrastructure resource management (IRM) monitoring, notification, logging, reporting, and correlation and analysis tools as shown on the right of Figure 5.3.

From a PCFE standpoint, metrics such as energy consumption, cost, heat generated, and CO_2 emissions need to be made available. IT resources need to be measured and looked at while considering many factors, includ-

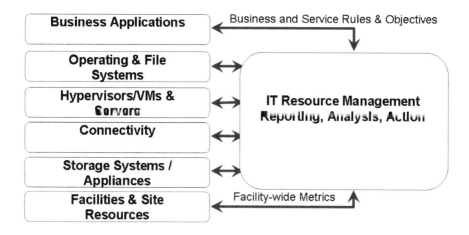

Figure 5.3   IT Resource and Service Points of Interest for Metrics and Measurement

ing activity or inactivity, environment, applications, compute or I/O centric, the amount of data being stored (storage centric) versus the amount being processed (server centric), and geographic location (which determines energy rates, availability and emissions).

A server-centric environment may use energy to power and cool servers compared to storage and other IT equipment. An environment that has a larger ratio of storage capacity and I/O operations with less compute or server resources may show a lower percentage of power used for servers and more for storage (disk and tape). Keep in mind that these are averages, and typical environments and "actual mileage will vary" depending on the factors specific to a situation.

Establishing a baseline set of measurements is important for many reasons, including establishing normal and abnormal behavior, identifying trends in usage or performance patterns, as well as for forecasting and planning purposes. For example, knowing the typical IOP rates and throughput rates for the storage devices, as well as the common error rates, average queue depths, and response times, will allow quick comparisons and decisions when problems or changes occur.

Baselines should be established for resource performance and response time, capacity or space utilization, availability and energy consumption. Baselines should also be determined for different application work scenarios in order to know, for example, how long certain tasks normally take. Baseline IRM functions, including database maintenance,

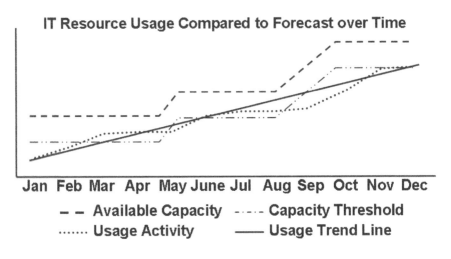

Figure 5.4   Measuring and Comparing IT Resource Usage over Time

backups, virus checking, and security scans, can be used to spot when a task is taking too long or finishing too fast, both of which could indicate a problem. Another example is that high or low CPU utilization compared to normal could indicate an application or device error resulting in excessive activity or blocking work from being done.

For planning purposes, resource usage and other key performance indicators can be plotted as shown in Figure 5.4 to show available resource capacity limits, thresholds for acceptable service delivery, actual usage, availability or performance, and trends. Establishing thresholds, which are usually less than the actual physical limits of the equipment, is useful for managing service delivery to a given response time or a particular performance or availability level.

For example, to meet a specific performance response time, servers or storage performances may be kept at less than full performance capacity. Or resource usage can be targeted, based on experience and historical baselines, below a certain percent utilization to ensure acceptable service delivery at a specific cost point.

From a forecasting and planning perspective, baseline comparisons can be used to determine or predict future resource usage needs while factoring in business and application growth. The benefit is that with a resource usage and performance capacity plan, the right amount and type of resources can be made available in a timely and cost-effective manner when needed. By

Table 5.1   Some Power, Cooling, Floor Space, and Environmental or IT-Related Metrics

Term	Description	Comments
AC	Alternating current	Type of electivity commonly used for most electrical or electronic devices.
Amps	Watts/volts	The flow rate of electricity, e.g., 144 watts/ 12 volts = 12 amps.
Annual kWh	kWh × 24 × 365	Amount of energy used in kilowatts in 1 year.
Bandwidth	Bytes per second	How much data is moved in 1 second. Used for measuring storage system as well as network data in-flight performance
Btu/hour	Watts × 3.413	Heat generated in 1 hour from using energy, in British thermal units. 12,000 Btu/hour equates to 1 ton of cooling.
CO_2 emission	On average, 1.341 lb per kilowatt-hour of electricity generated	The amount of average carbon dioxide (CO_2) emissions from generating an average kilowatt-hour of electricity.
DC	Direct current	Electivity used internally by IT equipment after conversion from AC.
Hz	Hertz	Frequency of electricity, such as 60 Hz (60 cycles per second) in North America or 50 Hz in other parts of the world.
IOPS	Input/output operations per second	Measure of the number of I/O operations, transactions, file requests, or activity in 1 second.
Joule	1 watt per second	Rate at which energy is used per second. One watt-hour is equivalent to 3,600 joules. 1 kWh represents 3,600,000 joules, or 1,000 joules per second.
kVA	Volts × amps/ 1,000 or kW/ power factor	Number of kilovolt-amperes.
kW	kVA × power factor	Efficiency of a piece of equipment's use of power.
Latency	Response time	How long work takes to be performed.

Table 5.1   Some Power, Cooling, Floor Space, and Environmental or IT-Related Metrics (continued)

Term	Description	Comments
MHz or GHz	Megahertz or gigahertz	An indicator of processor clock speed.
Power factor (pf)	Efficiency of power conversion	How effectively power is used and converted by power supplies. Power supplies that are 80% efficient or better may be referred to as "80-plus."
U or rack unit (RU)	1 U = 1.75 inches	EIA Electronic Industry Alliance (EIA) metric describing height of IT equipment in racks, units, or cabinets.
VAC	Volts AC	How many AC volts are being used.
Volt-amperes (VA)	Volts × amps	Power can be expressed in volt-amperes.
Volts or voltage	Watts/amps	The amount of force on electrons as a measurement of electricity (AC or DC).
Watt	Amps × volts or multiple Btu/hr × 0.293	Unit of electrical energy power to accomplish some amount of work.

combining capacity planning across servers, storage, networks, and facilities, different groups can keep each other informed on when and where resources will be needed to support server, storage, and networking growth.

Table 5.1 lists several metrics pertaining to electivity usage with descriptions and comments, including how to convert or translate a known metric to get an unknown value. For example, if you know the number of watts of power a device is using, you can determine the number of amps by dividing watts by known voltage. Similarly, if you do not know the number of watts for a device but you are given the number of Btu, simply divide the number of Btu by 0.293; for example, 1,000 Btu × 0.293 = 293 W. Metrics such as IOPS per watt are calculated by dividing the number of IOPS per second by the number of watts of energy used. Similarly, megahertz per watt or bandwidth per watt is found by dividing the number of megahertz or amount of bandwidth by the energy used.

Some metrics are measured and others are derived from measured metrics or are a combination of different metrics. For example, a storage system may report the number of I/O operations on a read and write basis along with the amount of data read and written. A derived metric is created by dividing bandwidth by number of I/O operations to get average I/O size. Similarly, if number of I/O operations and average I/O size are known, bandwidth can be determined by multiplying I/O rate by I/O size. Different solutions will report various metrics at different levels of detail. Similarly, third-party measurement and reporting tools, depending on data source and collection capabilities, will vary in the amount of detail that can be reported.

Bytes are counted using different schemes including binary base 2 and decimal base 10 (see Table 5.2). Networks traditionally have been measured in bits per second, whereas storage and related I/O are measured in bytes per second. These are sometimes referred to as "little b" (bits) and "big B" (bytes). Also shown in Table 5.2 are various abbreviations, including those of the international system of units (SI).

Intuitively, energy would be measured in terms of joules per second to parallel activity metrics per second. Generally speaking, however, electrical power is measured and reported on a kWh basis, in alignment with how utilities bill for power use. For example, if a device consumes 1,000 watts on a steady-state basis for 1 hour, it will consume 1 kWh of energy or 3,600,000 joules.

Metrics can be instantaneous burst or peak based, or they may be sustained over a period of time, with maximum, minimum, average, and

Table 5.2   IRM Counting and Number Schemes for Servers, Storage, and Networks

	Binary Number of Bytes	Decimal Number of Bytes	Abbreviations
Kilo	1,024	1,000	K, ki, kibi
Mega	1,048,576	1,000,000	M, Mi, bebi
Giga	1,073,741,824	1,000,000,000	G, Gi, gibi
Tera	1,099,511,627,776	1,000,000,000,000	T, Ti, tebi
Peta	1,125,899,906,842,620	1,000,000,000,000,000	P, Pi, pebi
Exa	1,152,921,504,606,850,000	1,000,000,000,000,000,000	E, Ei, exbi
Zetta	1,180,591,620,717,410,000,000	1,000,000,000,000,000,000,000	Z, Zi, zebi
Yotta	1,208,925,819,614,630,000,000,000	1,000,000,000,000,000,000,000,000	Y, Ui, yobi

standard deviation noted along with cumulative totals. These metrics can be recorded and reported by different time intervals, for example, by hour, work shift, day, week, month, or year.

IT technology manufacturers provide information about electrical energy consumption and/or heat (Btu/hour) generated under a given scenario. Some vendors provide more information, including worst-case and best-case consumption information, whereas others provide only basic maximum breaker size information. Metrics published by vendors and that should be visible on equipment may include kilowatt (kW), kV, amps, VAC, or BTU. Missing information can be determined if the other factors are available. For example, if a vendor supplies Btu/hour, the number of watts can be found by multiplying Btu/hour times 0.293. As an example, a device that produces 1,000 Btu/hour uses 293 watts.

Other metrics that can be obtained or calculated include those related to recycling, emissions, air flow, and temperature. Other metrics pertain to server CPU, memory, I/O and network utilization, capacity usage, and performance of local or internal storage. A compound metric is derived from multiple metrics or calculations. For example, IOPS per watt can be determined when base metrics such as IOPS and watts are known; these can be used to determine a metric for activity per energy consumed.

Application metrics include number of transactions, email messages, files, photos, videos, or other documents processed. Metrics for data protection include amount of data transferred in a given time frame, number of successful or failed backup or data protection tasks, how long different jobs or tasks take, as well as other error and activity information. Configuration management information includes how many of different types of servers, storage, networking components, software, and firmware, along with how they are configured.

These and other metrics can indicate a rate of usage as a count or percentage of a total such as server CPU measured from zero to 100% busy. Percent utilization gives a relative picture of the level of activity of a resource, percent utilization by itself does not indicate how service is being delivered or of the PCFE impact. For example, a server running at 50% utilization may consume less power than at 85%, yet, at 85%, application response time and performance may decrease in a nonlinear fashion. Performance of servers, storage, and networks typically degrade as more work is being done, hence the importance of looking at response time and latency as well as IOPS or bandwidth as well as percent utilization and space used.

5.3   Measuring Performance and Active Resource Usage

Generally speaking, as additional activity or application workload (including transactions or file accesses) is performed, I/O bottlenecks will cause increased response time or latency. With most performance metrics, such as throughput, the higher the value the better; however, with response time, the lower the latency the better. Figure 5.5 shows the impact as more work is performed (dotted curve): I/O bottlenecks increase and result in increased response time (solid curve). The specific acceptable response time threshold will vary by applications and service-level agreement requirements. As more workload is added to a system with existing I/O issues, response time will increase correspondingly (as shown in Figure 5.5). The more severe the bottleneck, the faster the response time will deteriorate. Eliminating bottlenecks allows more work to be performed while maintaining response time at acceptable service-level threshold limits.

To compensate for poor I/O performance and to counter the resulting negative impact to IT users, a common approach is to add more hardware to mask or move the problem. If overconfiguring to support peak workloads and prevent loss of business revenue, excess storage capacity must be managed throughout the nonpeak periods, adding to data center and management costs. The resulting ripple effect is that now more storage needs to be managed, including allocating storage network ports, configuring, tuning, and backing up data. Storage utilization well below 50% of available capacity is common. The solution is to address the problem rather than moving and hiding the bottleneck elsewhere (rather like sweeping dust under the rug).

Another common challenge and cause of I/O bottlenecks is seasonal and/or unplanned workload increases that result in application delays and frustrated customers. In Figure 5.6 a seasonal workload is shown, with seasonal spikes in activity (dotted curve). Figure 5.6 can represent servers processing transactions, file, video, or other activity-based work, networks moving data, or a storage system responding to read and write requests. The resulting impact to response time (solid curve) is shown in relation to a threshold line of acceptable response time. The threshold line is calculated based on experience or expected behavior and represents the level at which,

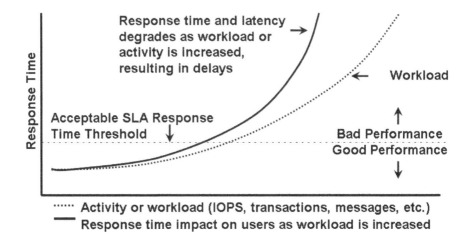

Figure 5.5   I/O Response Time Performance Impact

when work exceeds that point, corresponding response time will degrade below acceptable. For example, peaks due to holiday shopping exchanges appear in January, then drop off, and then increase again near Mother's Day in May.

Thresholds are also useful from a space capacity basis—for example, on servers, determining that particularly applications can run statistically at up to 75% utilization before response time and productivity suffer. Another example would be establishing that storage capacity utilization to meet performance needs for active data and storage is 70%, while near-line or offline storage can be utilized at a higher rate. It is important to note that the previous threshold examples are just that—examples—and actual thresholds will vary. Check application or system software as well as with hardware manufacturers for guideline and configuration rules of thumb.

For activity-based IT data center measurements, that is, where useful work is being done, activity per unit of energy metrics are applicable. In everyday life, a common example of activity or useful work per energy used is miles per gallon for automobiles or miles per gallon per passenger for mass transit including commercial aviation. Examples of data center useful work and activity include data being read or written, transactions or files being processed, videos or web pages served, or data being moved over local- or wide-area networks.

Activity per watt of energy consumed can also be thought of as the amount of work per energy used. A reciprocal is amount of energy per unit

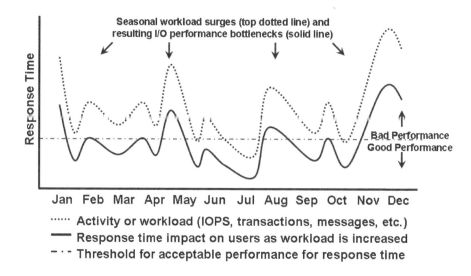

Figure 5.6   I/O Bottlenecks from Peak Workload Activity

of work performed. Activity per watt can be used to measure transient or flow-through networking and I/O activity between servers and storage devices, or between user workstations and an application server. Common examples of work per energy used are MHz per watt, IOPS, transactions, bandwidth or video streams per watt, storage capacity per watt, or miles per gallon. All indicate how much work is being done and how efficiently energy is being used to accomplish that work. This metric applies to active workloads or actively used and frequently accessed storage and data.

Figure 5.7 shows a simple example of eight servers and their average storage I/O activity. In the example, some servers are single attached, some are dual attached over different paths for redundancy to Fibre Channel or Ethernet storage. Activity-based measurements can occur at the server via third-party or operating system utility-based tools, via application-specific tools, or benchmark and workload simulation tools, via various tools.

As an example, assume that when server number 7 is actively doing work, the energy used is 800 watts, and that energy yields 800 IOPS (8 kbytes) or 1 IOPS per watt on average. The more IOPS per watt of energy used, the better. Of course, this example does not include networking or switch use, or storage system activity. If, say, the storage consumes on average 1,200 watts when doing active work, the number of kbytes per second from each of the servers can be totaled to determine bandwidth per watt of energy for the storage system. In this example, the storage system at average

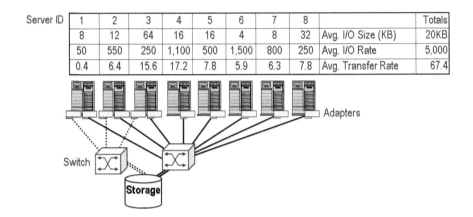

Server ID	1	2	3	4	5	6	7	8		Totals
	8	12	64	16	16	4	8	32	Avg. I/O Size (KB)	20KB
	50	550	250	1,100	500	1,500	800	250	Avg. I/O Rate	5,000
	0.4	6.4	15.6	17.2	7.8	5.9	6.3	7.8	Avg. Transfer Rate	67.4

Figure 5.7   Examples of Server, Storage, and Network Being Measured

Table 5.3   Activity Metrics Server, Storage, and Networking Example

Server	1	2	3	4	5	6	7	8	Total
Average watts per server	800	800	800	800	800	800	800	800	6,400
I/O size (kbytes)	8	12	64	16	16	4	8	32	
I/O rate per second	50	550	250	1,100	500	1,500	800	250	5,000
Bandwidth (kbytes per second)	400	6,600	16,000	17,600	8,000	6,000	6,400	8,000	69,000
IOPS per watt	0.1	0.7	0.3	1.4	0.6	1.9	1.0	0.3	
Bandwidth per watt	0.5	8.3	20.0	22.0	10.0	7.5	8.0	10.0	

active energy consumption of 1,200 watts yields about 57.5 kbytes per watt, or about 4.17 IOPS per watt.

For LAN and SAN networks, the process involves determining the amount of power used by a switch and then the amount being used per port. For example, the switches shown in Table 5.3 consume 1,600 watts each in total, but only 50 watts per port (32-port switch). Without factoring in activity (for example, the number of frames, packets, IOPS, or bandwidth) to support the previous example, 15 ports are needed yielding (15 × 50 watts) 750 watts.

Adding up the energy used for the servers, network, and storage along with activity or work being performed provides a more granular view of the

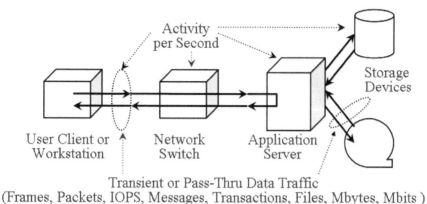

Transient or Pass-Thru Data Traffic
(Frames, Packets, IOPS, Messages, Transactions, Files, Mbytes, Mbits)

Figure 5.8   Bandwidth per Second for Transient Data In-Flight and Storage Systems

efficiency and productivity of the data center. Note that the above example does not consider whether the servers are fully utilized, what the response time or latency is per I/O operation, or what the time duration or period is. For example, the values shown in Table 5.3 may represent 24-hour averages, or they could represent a prime-time work shift. Also not factored into the above simple example is the amount of cooling required to remove heat, the density or footprint, or the usable capacity.

In Figure 5.8, at the application server, energy is used to perform work, including running applications, processing transactions, supporting email or exchange users, handling I/O operations, etc. On the far right, storage systems take in data written by servers and respond to read requests. Some of the data may be active, either being both read and written or updated, some of the data may be read-only or static reference or look-up data, whereas other data may be offline and inactive, including previous backup or archives for compliance or long-term data preservation.

Activity per watt can also be used to measure the amount of data moved to and from storage devices by servers or between other storage systems. Figure 5.8 shows activity being performed per unit of energy in several locations, including transactions being requested by a user or client workstation resulting in network traffic. In the middle of Figure 5.8, network traffic, including frames or packets per second, IOPS, transactions or bandwidth per second, passes through a switch that consumes energy to move data between the client on the left and the server on the right. Activity per watt is calculated as work (IOPS) divided by energy used (watts).

Note that when using an activity-per-energy measurement, it is important to know whether the energy consumed is just for the equipment or if it also includes cooling. Note also that activity per watt should be used in conjunction with another metric such as how much capacity is supported per watt in a given footprint. Total watts consumed should also be considered, along with price, to achieve a given level of service for comparative purposes.

Bandwidth per watt should not be confused with capacity per watt, such as terabytes of storage capacity space. This latter metric refers to the amount of data moved per second per energy used. Bandwidth per watt also applies to transient and flow-through networking traffic. It is also used for measuring the amount of data that can be read or written from a storage device or server, such as Mbytes per second per watt of energy for network traffic or switch.

IOPS per watt represents the number of I/O operations (read or write, random or sequential, small or large) per unit of energy in a given time frame for a given configuration and response time or latency level. Examples include SPEC and SPC-1 IOPS per watt or other workloads that report I/Os or IOPS.

Multiple metrics can and should be considered. It is important, however, to look at them in the context of how and where they are being used. For example, a fast solid-state disk (SSD) will have a high IOPS and low power consumption per physical footprint compared to a traditional disk drive. However, a SSD will usually also have a lower storage capacity and may require different data protection techniques compared to regular disk storage. Furthermore, while SSD may be more expensive on a capacity basis, on an IOPS-per-footprint basis, the SSD can have a lower cost per transaction, IOP, or activity performed than regular disk, albeit with less space capacity.

Exercise caution in combining too many different metrics in an attempt for a normalized view. The risk is that different metrics can cancel each other and lose their reflective value. An example of marketing magic is along the lines of (capacity * IOPS)/(watts * space). The challenge with marketing numbers is that their purpose is usually rooted in making sure that a particular vendor's technology looks more favorable than others rather than actual applicability.

For example, marketing may not factor in bandwidth versus I/O-intensive applications or the cost and applicable data-protection level. Another issue with "magic" metrics, like the one in the previous example, is that they can be mistakenly used to try and measure different types, tiers, and categories of storage. Another common mistake is to use raw disk drive numbers for theoretical performance, rather than combined storage controller and disk drive performance based on industry standard or other acceptable benchmarks. For true comparison purposes, look at applicable metrics relevant to the task and reflective of the function for which the resource will be used.

5.4   Measuring Capacity and Idle Resource Usage

Capacity metrics are important in data centers in many different technology domains. A variation of capacity-based measurements, briefly touched on previously, is idle or inactive resource usage. One scenario of idle usage is a server assigned to support an application that is inactive during certain periods. For example, a server is busy doing work for 12 hours of the day and then, other than supporting some IRM or data maintenance functions, is idle until the next day.

Anther variation is storage that is allocated and has data written to it for storage purposes, yet the data is not regularly accessed. An example is reference or static data, also known as persistent data, that does not change but is written to disk and seldom, if ever, read. To help optimize performance, availability, capacity, and energy use, inactive data can be moved to offline storage such as magnetic tape. Once written to tape, the storage does not require power to be stored. Some disk drive-based solutions are replacing or coexisting with tape and optical-based storage solutions for near-line storage of inactive data. For online and active disk storage, performance is an important metric, along with availability, capacity, and energy use; near-line and offline disk based storage has a lesser focus on performance.

For space capacity-based storage and, in particular, storage for idle data including backup targets or archiving solutions involving data de-duplication, there is a tendency to measure in terms of de-duplication ratios. Ratios are a nice indicator of how, for a given set of data, the data and its footprint can be reduced. Data movement or ingestion and processing rate, which is the rate at which data can be reduced, is a corollary metric for data-reduction ratios. Data-reduction rates, including compression rates, can indicate how much data can be reduced in a given window or time frame.

The amount of equipment per rack or the number of cabinets per footprint reflects IT resource density in a given footprint. This is a compound metric that can look at the total power consumption of all equipment installed in a given rack or cabinet footprint. The metric can also factor in the amount of resource space and capacity available or the number of servers or storage devices. A caveat in using this type of metric is to be sure to include the height of the rack or equipment cabinets. Variations of this metric can be the amount of resources, including performance IOPS or bandwidth, capacity, data protection, number of servers or network ports, per "U" or rack unit. Another variation is to base the footprint on floor space in square feet or meters while keeping in mind that some equipment utilizes deep, wide, or other nonstandard cabinets. Metrics can be obtained by measuring intelligent power distribution units and power management modules in the cabinets or from the individual components in a rack or cabinet. Another factor with regard to footprint is how much weight in addition to power and cooling is required or supported, particularly when shifting equipment.

Still another metric variation looks at the amount of storage capacity per watt in a given footprint. This is useful for inactive and idle storage. This metric is commonly used by vendors, similar to how dollar per capacity ($/GB) is often used, for comparing different storage technologies. One issue with this metric is whether it is considering the capacity as raw (no data protection configuration such as mirroring, no file system or volume formatting) or as allocated to a file system or as free versus used. Another issue with this metric by itself is that it does not reflect activity or application performance or effectiveness of energy per unit of capacity to support a given amount of work—for example, watt per tape cartridge when the tape is on a shelf versus when the tape is being written to or read. Another concern is how to account for hot spare disk drives in storage arrays. Also, the metric should account for data offline as well as data online and in use.

A point of confusion can exist around the use of Mbyte, Gbyte, Tbyte or Pbyte per watt when the context of bandwidth or capacity is missing. For example, in the context of bandwidth, 1.5 Tbyte per watt means that 1.5 Tbytes per second are moved at a given workload and service level. On the other hand, in the context of storage space capacity, 1.5 Tbyte per watt means that 1.5 Tbytes are stored in a given footprint and configuration. The takeaway is not to confuse use of bandwidth or data movement with storage space capacity when looking at Tbyte or Pbyte and related metrics per watt of energy.

5.5   Measuring Availability, Reliability, and Serviceability

Availability can be measured and reported on an individual component basis, as a sum of all components, or a composite of both. A balanced view of availability is to look at the big picture in terms of end-to-end or total availability. This is the view that is seen by users of the services supported by the storage network and its applications.

The annual failure rate (AFR) is the association between MTBF and the number of hours a device is run per year. AFR can take into consideration different sample sizes and time in use for a device. For example, a large sample pool of 1,000,000 disk drives that operates 7 × 24 hours a day (8,760 hours a year) with 1,000 failures has an AFR of 8.76%. If another group of similar devices is used only 10 hours a day, 5 days a week, 52 weeks a year (2,600 hours) with the same sample size and number of failures, the AFR is 2.6%. MTBF can be calculated from AFR by dividing the total annual time a device is in use by the AFR. For the previous example, MTBF = 2,600/2.6 = 1,000. AFR is useful for looking at various size samples over time while factoring in duty or time in use for availability comparison or other purposes.

There can be a cost associated with availability that needs to be understood to determine availability objectives. Vendors utilize terms such as five nines, six nines, or higher to describe their solution's availability. It is important to understand that availability is the sum of all components and their configuration. Seconds of downtime per year is calculated as 100% × [(100 N)/100]. where N is the desired number of 9's of availability as shown in Table 5.4. Availability is the sum of all components combined with design for fault isolation and containment. How much availability you need and can afford will be a function of your environment, application and business requirements, and objectives.

Availability is only as good as the weakest link in a chain. In the case of a data center, that weakest link could be the applications, software, servers, storage, network, facilities and processes, or best practices. Virtual data centers rely on physical resources to function; a good design can help eliminate unplanned outages to compensate for individual component failures. A good

Table 5.4   Availability Expressed as Number of 9's

Availability (%)	Number of 9's	Amount of downtime per year
99	0	3.65 days/year
99.9	1	8.77 hours/year
99.99	2	52.6 minutes/year
99.999	3	5.26 minutes/year
99.9999	4	31.56 seconds/year
99.99999	5	3.16 seconds/year
99.999999	6	∫ second/year

design removes complexity while providing scalability, stability, ease of management, and maintenance as well as fault containment and isolation.

5.6   Applying Various Metrics and Measurements

The importance of these numbers and metrics is to focus on the larger impact of a piece of IT equipment including its cost and energy consumption, and factoring in cooling and other hosting or site environmental costs. At a macro or "big-picture" level, the energy efficiency of an IT data center can be measured as a ratio of power being used for HVAC (heating, ventilating, and air conditioning) or cooling and power for IT equipment independent of actual work being done or data being stored. An example is the metric called **power usage effectiveness (PUE),** and its reciprocal, **data center efficiency (DCiE),** being put forth by the Greengrid industry trade group.

PUE is defined as the total facility power divided by total IT equipment (excluding HVAC). This provides a gauge of how effectively electrical power is being used to power IT equipment. DCiE is the reciprocal PUE and is defined as 1/PUE or total IT equipment power consumption divided by total facility power (including HVAC) × 100%. PUE can range from 1 (100% efficiency) to infinity. For example, a PUE of 3, the equivalent of a DCiE of 33%, indicates that about 70% of power being consumed by a data center is used for non-IT equipment.

Another metric being put forth by the Greengrid is **Data Center Productivity (DCP),** which considers active or useful work being done per energy being consumed. DCP is a very generic and broad metric that will take time to evolve and define what determines work or activity across different types, tiers, and categories of servers, storage, and networking devices. In the near term, a subset example of DCP could be considered to be activity per watt of power.

Another metric or series of metrics attempts to look at the total impact of the emissions and environmental health and safety footprint over the entire life of an IT resource. These could have focus on only the time frame an IT organization owns and operates the equipment, or they could be cradle-to-grave metrics that also factor in vendor manufacturing and materials, green supply chain and ecosystem, through acquisition, installation, and use until final disposition and recycling.

Questions that arise with an emissions metric is whether it is only for CO_2 and emissions-related tax accounting purposes or whether it is designed to provide a complete picture of the impact on environmental health and safety. Does the footprint or impact include associated IRM tasks and overhead, including data protection, backup, business continuity, disaster recovery and associated software running on different servers? Other issues to consider include whether the emissions footprint is based on averages or on a specific location and energy fuel source.

A simple emissions footprint can be determined by multiplying the number of watts used for a device plus applicable watts associated with cooling the equipment times 1.341 per kWh used to derive the number of CO_2 tons/kWh of electricity used. For example, if a device consumes 1,200 watts plus 600 watts (1.8 kWh) for cooling, the average hourly average emissions is 2.413.8 lb of CO_2 or about 10.57 tons of CO_2 per year. Note that 1.341 lb/kWh is an average and will vary by location, depending on energy costs and fuel source. Another note is that this simple calculation does not take into consideration the actual amount of carbon, which may be taxed separately under some emission tax schemes.

5.7 Sources for Metrics, Benchmarks, and Simulation Tools

Metrics and measurements can be obtained from many different sources, including external probes or analyzers, built-in reporting facilities, and add-on third-party data collection and reporting tools. Sources for information include event, activity, and transaction logs and journals as well as operation system and application-based tools. Vendors of servers, storage, and networking solutions also provide varying degrees of data collection and reporting capabilities. Metrics can also be obtained from facilities monitoring tools, power conversion, and distribution units, computer room air- conditioners, and other points within the data center, including security and event logs or alerts. For example, temperature and air flow data can be used to determine conflicts between hot and cold zones or areas of confluence, or how much energy is being consumed and how effectively it is being converted.

IT equipment power, cooling, footprint, and environmental metrics, including basic power requirements, can be found on equipment labels or nameplates Other sources of equipment PCFE metrics include vendor documentation such as site installation and planning guides or vendor websites that may also include sizing and configuration tools or calculators. Note that equipment nameplate values may exceed the actual power and cooling requirements for the equipment, as these values have a built-in safety margin; for a more accurate assessment, review vendor specification documents.

Performance testing and benchmarks have different meanings and different areas of focus. There is testing for compatibility and interoperability of components. There is performance testing of individual components as well as testing of combined solutions. Testing can be very rigorous and thorough, perhaps beyond real-world conditions; or testing can be relatively simple, to verify data movement and integrity. What is the best test for your environment depends on your needs and requirements. The best test is one that adequately reflects your environment and the applications' workload and can be easily reproduced.

5.8 Summary

Virtual data centers require physical resources to function efficiently and in a green or environmentally friendly manner. Thus it is vital to understand the value of resource performance, availability, capacity, and energy (PACE) usage to deliver various IT services. Understanding the relationship between different resources and how they are used is important to gauge improvement and productivity as well as data center efficiency. For example, while the cost per raw terabyte may seem relatively inexpensive, the cost for I/O response time performance needs to be considered for active data.

Having enough resources to support business and application needs is essential to a resilient storage network. Without adequate storage and storage networking resources, availability and performance can be negatively impacted. Poor metrics and information can lead to poor decisions and management. Establish availability, performance, response time, and other objectives to gauge and measure performance of the end-to-end storage and storage–networking infrastructure. Be practical, as it can be easy to get caught up in the details and lose sight of the bigger picture and objectives.

Additional key points include:

- Balance PACE and PCFE requirements to various levels of service.

- Compare resources "apples to apples," not "apples to oranges."

- Look at multiple metrics to get a multidimensional view of resource usage.

- Use caution in applying marketing-focused "magic" metrics that may not reflect reality.

- Metrics and measurements can be obtained and derived from many different sources.

- Use metrics and link to business and application activity to determine resource efficiency.

- Establish baseline metrics and profiles to compare current use with historical trends.

Additional companion material about data infrastructure metrics, performance, and capacity planning can be found on the companion website, www.thegreenandvirtualdatacenter.com and in Chapter 10 of my book, *Resilient Storage Networks—Designing Flexible Scalable Data Infrastructures* (Elsevier, 2004).

Chapter 6

Highly Effective Data Center Facilities and Habitats for Technology

Green Acres Is the Place to Be—Popular syndicated TV show

In this chapter you will learn:

- How data centers use and provide power and cooling for IT equipment
- Standby and alternative power and cooling options for data centers
- Environmental health and safety issues and relevant standards

New IT data center facilities can be designed to be green and energy efficient. Existing facilities can be upgraded and enhanced to address various energy and environmental concerns. It is critical to understand the various components and practices that comprise a physical data center. Beyond the basics, abstracted items that make up cloud and managed-service virtual data centers must be accounted for to implement strategies to achieve a green and virtual data center.

As mentioned in previous chapters, virtual data centers require physical resources to function. Similarly, physical IT resources, including servers, storage, and networks, require habitats or physical facilities to safely and securely house them. An IT data center is similar to a factory in that it is a combination of equipment used to transform resources into useful goods and services.

A factory is more than a building; it also requires equipment and raw resources to be processed. And a factory is more than a room or building full of equipment in which to process raw goods. A productive factory requires energy, management, automation, and other tools to run at peak efficiency. In the case of IT data centers, best practices, metrics, and measurements for insight into **infrastructure resource management (IRM)** combined with facilities and equipment operating in an energy-efficient manner result in a viable and economical data center.

6.1   Data Center Challenges and Issues

As information factories, IT data centers continue to evolve to support existing requirements as well as to provide additional services and capabilities to meet changing business and market demands. Chapter 1 talked about the green gap that exists between green perception and common data center **power, cooling, footprint, and environmental (PCFE)** issues. Closing or eliminating the green gap involves understanding the many facets of being green as a way of doing business, including actually addressing PCFE issues as opposed to being preoccupied with being perceived as green. Given that data centers provide a safe and secure habitat, including power, cooling, and floor space, for IT resources, addressing the green gap involves taking a closer look at various data center challenges and issues.

All data centers, whether company-owned physical facilities, a virtual out-sourced or managed service facility, or a cloud-based service, rely on some real and physical facility somewhere. Likewise, Web and traditional applications that rely on virtual servers, virtual storage, and virtual networks also need physical servers, storage, and networks housed somewhere.

Understanding the importance of physical IT resources requires understanding and insight into the various interdependencies that support and sustain virtual environments. In general, IT has gone from the situation of several decades ago, which meant fine-tuning scarce and expensive server, storage, and networking technology for configuration, performance, and application, to a period in which hardware is cheap, allowing the possibility of simply throwing more hardware at a problem to resolve it.

Neither approach is intrinsically good or bad, but with an increased focus on abstraction and moving applications and their developers farther away from hardware or even underlying software dependencies, productivity increases. With this boost in productivity must come a boost in the efficiency of how IT resources are used and made available to support applications that require more processing power, storage, and input/output (I/O) capacity per business function. For example, simply throwing more hardware at a business application, software, or hardware problem may not be practical given PCFE constraints. While hardware may be cheap, the costs to house, power, cool, and manage that hardware, along with associated software costs, are not cheap and in fact continue to rise.

In addition to supporting growth of servers, storage and networks, data centers need to:

- Support service-oriented infrastructure and applications
- Sustain business growth and dynamic market conditions
- Enhance business agility and IT service delivery
- Reduce or contain costs by doing more with less
- Boost efficiency and productivity of IT equipment and PCFE resources
- Address PCFE issues and challenges
- Increase amount of resources managed per person and IT budget dollar

Common data center issues and challenges include the continued demand for more IT services, including processing, storage, and I/O to support more data being stored with more copies being retained for longer periods of time. Compounding these are the rising price of energy, limits on power availability, and constrained electricity distribution networks. In current and older-generation data centers, half of all power is consumed by HVAC (heating, ventilating, and air conditioning), and power distribution depends on the efficiency of existing air-handling units, chillers, and related HVAC equipment. In addition to electrical power and cooling challenges, lack of floor space due to equipment sprawl is common, particularly in older facilities or where newer, denser IT equipment has yet to be installed.

Other common challenges faced by IT data centers include:

- Overcooling facilities based on dated best practices or older equipment standards
- Increasing density of IT equipment occupying reclaimed floor space and available power

- Constraints on existing standby power-generation or cooling capabilities
- Existing and emerging recycling and equipment disposition legislation
- Emerging emissions tax schemes and energy efficiency regulations
- Shrinking or frozen capital and operating budgets, and the need to do more with less

The increasing density of IT equipment to sustain business growth and support new application capabilities is also resulting in several challenges. Challenges and characteristics of higher-density IT equipment include:

- More power required per square or cubic foot (or meter)
- More servers, storage, and networking equipment per rack or cabinet
- More server compute, storage capacity, and performance per footprint
- More weight per footprint, requiring stronger floors
- More cooling required per footprint from denser equipment
- More dependence on resources in a given footprint, requiring high availability
- More images, virtual machines, or operating systems per physical server per footprint

Power consumption per footprint varies widely by business focus and types of applications, the ratio of servers to storage, the amount of online active storage compared to near-line or static and inactive data or networking bandwidth. For example, a compute-intensive entertainment video rendering, modeling, or simulation-based environment may have a higher ratio of servers to storage, an online hosting environment may have a balance of servers and storage, and a fixed or static content provider, or managed backup service, may have a higher ratio of storage to servers.

Density requires a balancing act to enable dense IT equipment to support growth without aggravating or further compounding PCFE issues. In

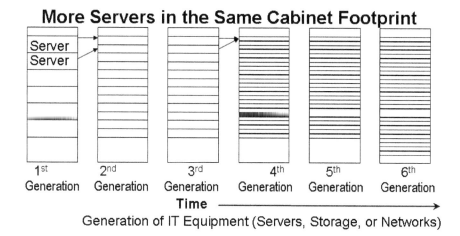

Figure 6.1   Increasing Density over Time of IT Equipment Generations

Figure 6.1, several generations of IT equipment are shown occupying the same physical footprint (floor space). In addition to fitting within the same footprint, the various generations of servers also have to exist within the available primary and secondary (standby) power as well as the available cooling footprint. In Figure 6.1, each successive generation of servers is shown as able to fit into a smaller physical footprint.

In addition to occupying less physical space within a cabinet, rack, or blade center, each successive generation also provides an increase in net processing or compute power along with a boost in memory and I/O capabilities. Figure 6.2 shows an example of how primary and secondary power constraints exist until a future upgrade to facility powering capabilities boosts the per-cabinet power footprint. Also shown in Figure 6.2 are the relative increases in power required per cabinet as the quantity of servers, their speed, and the number of processing cores increase.

In Figure 6.2, the required power per server does not scale linearly with the increase, as some improvement in energy efficiency and power management are being factored in to offset the growth. An important takeaway from Figure 6.2 is how capacity planning for servers, storage, and networks needs to incorporate the facility's PCFE capabilities to address how technology improvements can help sustain growth in existing footprints. Likewise, facilities planning should include server, storage, and networking capacity growth plans for near-term and long-term strategic planning purposes. The net result is that more applications can be supported and more information

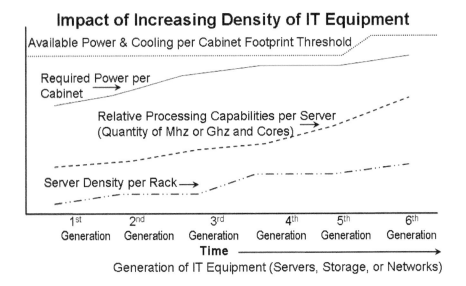

Figure 6.2   Relative Footprints and Resource Capabilities over Time

processed and stored, allowing the business to continue to grow and remain competitive with a more efficient and productive data center.

Server consolidation using virtualization is part of an approach to addressing PCFE issues. Consolidation by itself only addresses those servers and applications that lend themselves to being consolidated; it does not help or address applications that need to scale beyond the limits of a single or even multiple servers. Thus, in addition to server consolidation for candidate servers and applications, deployment of newer-generation, faster, and more energy-efficient equipment helps to address various PCFE issues.

The good news that comes with increased density is that servers will continue to become more powerful, as they have done over the past couple of decades, while fitting into a smaller physical footprint and requiring the same or less power while providing larger memory capacity. Likewise, storage systems that leverage smaller-form-factor, high-performance or high-capacity hard disk drives that consume less power and weigh less continue to be packed into denser footprints.

The bad news is that power, cooling, and floor space that may be reclaimed or become surplus with denser equipment will be absorbed by growth. Also with density comes new challenges, including more power and cooling required per footprint, cabinet, or rack than in the past. The net result, however, is that recapturing footprint in terms of PCFE capabilities in order to support expansion of IT resources is an important step to be taken by IT organizations to sustain business growth.

6.2   What Makes up a Data Center

IT data centers vary in shape and size, focus, and design approach reflecting various preferences and technology options. Some data centers or technology habitats are very small—for example, an equipment closet or room—whereas others are ultralarge, spanning the equivalent of several football fields in size. Depending on the types of applications being hosted, some data centers have more servers than storage or networking equipment; others have a higher ratio of storage to servers and networking. A networking-centric or telecom-related facility will tend to have more networking-related equipment; however, given the rise in demand for video and enhanced network services, more servers and storage devices are appearing in network and telecom facilities. Other variations include the number of different rooms or zones for equipment isolated from each other in the same or adjacent buildings. Data centers can exist in purpose-built facilities above or below ground or in retrofitted existing facilities.

Basic or essential components of a habitat for IT equipment resources include:

- Electrical power—primary and standby, distribution and conversion
- Cooling—chilling, heating, ventilation, air conditioning (HVAC)
- Floor space and cable management—raised floor or overhead conveyance
- Practices and management related to environmental health and safety—recycling, reuse, and disposal of resources

Additional considerations include:

- Fire detection and suppression
- Computer room air conditioners (CRACs)
- Smoke and exhaust ventilation and air filtration

- Physical security, monitoring, and surveillance
- Receiving, shipping, storage, and staging areas
- Workspace and operations control centers
- Mechanical rooms, zones
- Water and fuel reserves for standby power
- Communications write closets and demarcation points
- Racks, cabinets, cable management and conveyance
- Asset management, monitoring, measurement, test, diagnostic, and IRM tools

6.2.1   Tiered Data Centers

Several tiers of data centers are possible, with tier 1 being lowest availability and cost to implement and tier 4 being highest level of protection and most expensive habitat for technology. Note that within a given facility, different rooms or zones can have different tiers of protection. For example, a tier 4 data center might have some rooms or zones configured for tier 4 with others being tier 3, tier 2, or even tier 1 for lower-cost, basic services for applications that do not need high availability. Various data center standards including TIA-942, the standard of the Telecommunications Industry Association, can be found at www.tiaonline.org.

6.2.1.1   Tier 1 Data Center

A tier 1 data center, equipment room, or closet is a basic habitat for technology without redundant equipment, utility, or network services. Characteristics of a tier 1 habitat for technology include possible single points of failure or a lack of fault isolation design. Availability can be improved by adding UPS (uninterruptible power supplies) or standby generators and secondary telecommunications and network capabilities. Some level of data protection, including basic backup either onsite or offsite, leveraging a managed service provider, may be in place. Organizations or applications that use tier 1 facilities include small businesses or **small office/home office (SOHO)** locations that can tolerate some disruptions or downtime, which may rely on a managed service provider or hosting service for online presence and Web-based services.

6.2.1.2   Tier 2 Data Center

A tier 2 data center or room within a larger data center combines features of tier 1 plus redundant IT equipment to improve availability and data protection. Characteristics of tier 2 centers include multiple servers, storage, and networking devices, telecommunications that are vital to businesses that are dependent on email, Web, and other online services and presences. A tier 2 environment includes a formal data center or equipment room that is separate from other areas. Basic redundancy in power and cooling include some form of UPS and a standby generator capable of handling all or some of the power load for 24 hours. Applications and data are covered by some form or level of business continuity and disaster recovery protection. Physical site selection is not as critical as for higher-tier data centers that provide additional availability, protection, and security.

6.2.1.3   Tier 3 Data Center

A tier 3 data center or room within a larger data center combines the features of a tier 1 and a tier 2 data center with the addition of redundant powered equipment and networks to improve availability, accessibility, and survivability. Additional characteristics of a tier 3 data center include redundant primary utility paths and sources, either active passive or both active (load sharing), redundant power and cooling, redundant network bandwidth service providers, and a secondary or alternate site (active or standby) for business continuity or disaster recovery protection. Careful site selection that considers proximity to applicable threat risks, available power and networking connection points, cost, and transparency is also important.

For further resiliency, tier 3 data centers support tolerance of 1-hour fire rating with applicable detection and suppression for fire and smoke. They support concurrent maintenance without disrupting applications availability or performance. They have a high level of security, with the ability to tolerate a 72-hour power outage with adequate fuel and water reserves. The physical structure is somewhat hardened for applicable threat risks. Tier 3 candidate applications or organizations include those with a worldwide presence, highly dependent on IT and associated servers, storage, and networking resources being available 7 × 24 × 365. They have minimal to no tolerance for unscheduled service disruptions because of the high cost of downtime to business revenue, delivered service, image, or a combination of all of these.

6.2.1.4   Tier 4 Data Center

Tier 4 data centers combine tier 1 plus tier 2 plus tier 3 features and add fully redundant and fault-tolerant equipment and availability capabilities. Additional characteristics of a tier 4 data center or habitat for technology include separate independent and isolated utility paths and independent network paths from divergent sources. Other characteristics include redundant power (primary and secondary) and cooling, careful site selection, alternative sites for business continuity or disaster recovery, a minimum of 2-hour fire rating, a high level of physical security, and ability to tolerate 96 hours of power outage. Tier 4 centers have around-the-clock onsite (or in adjacent site) electronic monitoring and security surveillance. There is isolation of different rooms and zones between shipping and receiving, staging, vendor-secured areas, and main floor space or compartmentalized zones. The physical facility is hardened to applicable threat risks including tornadoes, hurricanes, floods, or other acts of nature.

Organizations or applications that require tier 4 data center protection include large businesses, high-profile businesses, and businesses that are highly dependent on IT equipment and services being available. Applications that rely on tier 4 data centers can be independent of business revenue size or number of employees. The common theme for applications and businesses requiring tier 4 capabilities are the high cost of downtime in terms of lost revenue, lost productivity, spoilage or loss of raw goods, lost opportunity, public image and perception, regulatory fines, or support for essential services including emergency dispatch, among others.

6.3   Data Center Electrical Power and Energy Management

IT data centers require electricity to operate servers, storage, and networks as well as associated cooling and ventilation equipment. Energy usage may not be well understood by users or even by most IT professionals or vendors. Better information about energy use is needed, besides what is provided on equipment nameplates, for effective energy allocation and configuration. Data on equipment nameplates may provide for growth without clearly stating what energy is immediately required and what is included for growth, resulting in overconfiguration. However, if electrical power is overestimated, subsequent cooling capabilities may also be over-configured, resulting in excessive cooling and energy consumption.

Electrical power is generated, transmitted, and used as either alternating current (AC) or direct current (DC). Utility-provided electricity is AC. Most commercial and consumer electronics operate on supplied AC power that is converted to DC, if necessary, internally. For example, servers and storage devices are supplied with AC power that is then converted to DC by internal transformers. Items that operate off battery power either use DC natively or require conversion using a DC-to-AC inverter. AC electrical power voltages may be supplied as 120, 208, 240, or 480 volts, depending on the type of service and number of phases (one, two, or three). Frequency varies with location: North America uses 60 hertz (Hz), whereas Europe and other parts of the world use 50 Hz.

Electricity is referenced in kilowatt-hours (kWh), such as on an energy bill, or in kilovolts AC (kVA). For example, a small 12-kW generator provides 50 amps at 240 volts or 12 kVa at 60 Hz frequency. Power is sometimes reported in kilovolt amperes (kVa), which is kilovolts or thousands of volts multiplied by amps. For example, single-phase 208 volts at 1,000 amps is (0.208 kV × 1,000 amps) = 208 kVa. KVa can also be determined by dividing known kilowatts by a known power factor(pf) or by 0.8 if the power factor is not known. KVa may be listed on equipment nameplates, site planning and installation guides, or associated documentation.

Figure 6.3 shows an example of how power is supplied by electric utilities from generation sources or power plants via transmission networks to data centers. Depending on the size and importance of applications being hosted in a data center, multiple feeds from different power sources or access points to the broader electric power grid may exist. The feed from the electric utility is considered a primary feed, with secondary or standby power provided by on-site generators.

To maintain availability during a power outage and the short time before generators start up and are ready to provide stable power, uninterruptible power supplies (UPS) are used. Power management switches enable automatic or manual transfer from primary utility or self-generated power to standby power for maintenance, diagnostics, and troubleshooting, or for maintenance bypass with load balancing. Power distribution units (PDUs) transform higher-voltage power supplied from energy management systems to IT equipment where it is needed. Electrical power cables can be found either under raised floors or in overhead conveyance systems. For example, PDUs can be located throughout a data center in different zones, providing

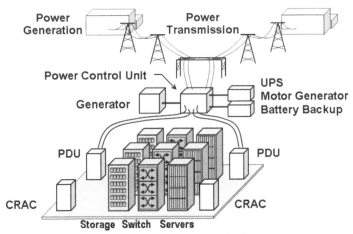

Uninterruptible Power System (UPS), Power Distribution Unit (PDU), Computer Room Air Conditioning (CRAC)

Figure 6.3   Powering a Data Center or Habitat for Technology

power to servers, storage, and networking devices along with HVAC and computer room air conditioning (CRAC) units.

In Figure 6.3, a power control unit, power switch, or energy management system is fed power by one or more utility feeds. Also attached to the power control unit is a UPS system that provides short-term power using batteries or a motor generator until standby generators are ready for use. A motor generator draws power from the utility to turn a large flywheel that, in the event of a loss of primary power, provides enough kinetic energy to turn a small generator long enough for gas, diesel, propane, or natural gas generators to start up and stabilize. An alternative, or, for extra redundancy, in addition to a motor generator, is a battery-based backup system that, depending on the size and number of batteries, can provide standby power for the data center.

6.3.1   Secondary and Standby Power

Generators can be used in continuous operation mode for self-generation or co-generation of power to supplement utility-provided electrical power as well as for standby power in the event of a power outage. Self- or co-generation is an option to supplement power demands during peak energy periods, when utility-provided power may be more expensive and subject to availability. Another use of self-generated power is to proactively reduce or turn off utility-provided power during peak periods in exchange for incentives, including rebates or reduced off-hour energy pricing. Generators range in capacity from a few kilowatts, for home or personal use, to hundreds of kilowatts or megawatts. Fuel options include gas, propane, natural gas, and diesel. Depending on the fuel source used, local or on-premises holding tanks will be needed to hold reserves for extended operation.

UPS systems provide a bridge to cover the gap between the time utility-supplied power is disrupted and the time a generator is started and able to produce clean power. Depending on the size and type of generator, as well as the power management or control system, the generator may start automatically when the control unit detects loss of power. The generator may start up within a few seconds plus another few seconds for the power to stabilize for use. During the gap between loss of utility power and stable power from the generator, either a battery-powered UPS or a motor generator whose flywheel continues to rotate from kinetic energy provides bridge power to keep IT equipment running.

Online UPS provide continuous clean (no surges or fluctuations) power to equipment with no break when transferring from the utility or primary source to standby battery power, providing protection from blackouts, brownouts, surges, and transient power fluctuations. A line interactive UPS provides power conditioning and filtering, though with a short (several milliseconds) disruption when transferring from primary or utility power to batteries. This short disruption is typically tolerable to equipment power supplies. Offline UPS pass utility power through to protected equipment without any filtering; usually transferring to battery protection within a few milliseconds. Small, self-contained and affordable battery-powered UPS of various sizes should be used for sensitive equipment in SOHOs as well as digital homes.

For large data centers, large rooms with banks of batteries may be needed to provide temporary bridge power until generators become ready for use. The caveat with UPS systems is that their batteries must be maintained and eventually disposed of in an environmentally friendly means in compliance with various regulations. Batteries need to be maintained, including periodically adding distilled water, cleaning, and charging. Batteries also require rooms for safe storage and containment away from other equipment. Table 6.1 provides an example of standby and alternate power options for various-sized data centers and businesses.

6.3.2   Alternative Energy Options and DC Power

Although most IT equipment uses AC power, some equipment, including network switches or specialized servers and storage devices in telecomm or rugged environments, operates on DC power. The requirement for DC power in telecom switching facilities stems from the need to operate on batteries. Equipment installed in telecom facilities generally adheres to the Network Equipment—Building System (NEBS) standards, of which there are many different levels and requirements. One feature of the NEBS system is the use of -48 volt DC power.

In an effort to optimize electrical power usage, some manufacturers and data centers are exploring use of DC powered servers on a large scale for large and dense servers. The premise is to reduce the number of AC-to-DC converters that are needed. The caveat is that simply installing a DC-powered large server or other piece of IT equipment may be easier said than done with current data center power equipment. However, moving forward, con-

Table 6.1   Examples of Standby and Alternate Power Options for Various-Sized Data Centers

	Standby Generator	Alternate Power Feed	Battery Backup or UPS	Automatic Power Switching	Surge Suppression	Perform an Energy Assessment
Ultralarge data center	Yes	Yes	Yes	Yes	Yes	Yes
Large data center	Yes	Yes	Yes	Yes	Yes	Yes
Medium-size data center	Yes	Yes	Yes	Yes	Yes	Yes
Small data center	Yes	Yes	Yes	Yes	Yes	Yes
Equipment closet	Optional	Optional	Optional	Optional	Yes	Yes
Coffee, copy room	Optional	Optional	Optional	Optional	Yes	Yes
SOHO	Optional	Optional	Optional	Optional	Yes	Yes
Digital home	Optional	Optional	Optional	Optional	Yes	Yes

sidering DC power distribution during planning for remodeling, expansion, or new data center building projects is worth considering.

Electric utilities are leveraging many different types of energy and fuel sources, including coal, nuclear, natural gas, oil, wind, hydro, solar, methane, and fuel cell-based solutions. In some areas, such as the Columbia River Valley, Las Vegas and the Hoover Dam region, the Tennessee River Valley, Norway, and China's Three Gorges Dam, where there is an abundance of flowing water, hydro-power electricity is produced. In locations with an abundance of open space and frequent wind activity, wind-generated electricity is an option. Solar panels are another alternative energy source for locations and facilities, either as supplemental power or as a portion of primary power. Solar panels, for example, can be installed to supplement traditional power generation for charging batteries in UPS systems in addition to providing some regular energy when the sun is shining.

Fuel cells are an emerging energy-generation technology that can be used for a wide variety of fixed and movable applications. A fuel cell has two electrodes, a positive anode and a negative cathode. A fuel cell generates electricity by a chemical reaction as a result of electrons transferring charged particles from one electrode to the other, using a catalyst to speed up the reaction. Hydrogen is a basic fuel used along with oxygen, producing water as a by-product. A single fuel cell can generate only a small amount of power; consequently, multiple cells usually need to be linked together. For data centers that plan to adopt DC-based power distribution for all or some equipment, a benefit of local fuel cells is the elimination of DC-to-AC-to-DC power conversion.

The benefits of fuel cells as an energy source include efficient use of fuel and virtually no pollution in the form of greenhouse gas emissions such as NO_2, CO_2, or SO_2 compared to traditional fossil fuel energy generation. Other benefits of fuel cell-based energy generation include quiet operating modes and the economy of operating by using fuel more efficiently. Using fuel cell-based electric power generation also can qualify a data center for energy efficiency certificates, rebates, and other incentives. There are many types of fuel cells, ranging from small portable devices which can be placed into an automobile or other mobile applications to fixed power plants. Fixed power plants or generation facilities can produce both heat and power from various fuel sources including methane, natural gas, propane, and hydrogen.

Unlike wind or solar energy generation, which depend on available wind and clear sunny days, respectively, fuel cells can have a much higher availability as a standby or primary energy source. For example, moving forward, a data center facility can leverage primary utility power when it is readily available and switch to co-generation or self-generation using fuel cells when appropriate. Fuel cells, like solar and other alternative energy sources, can also be used for charging standby batteries for UPS systems, off-loading energy demand from primary energy sources. Fuel cells are still an evolving technology, but as they mature and become more affordable they should be able to provide primary and secondary power for data centers.

6.4   Cooling, HVAC, Smoke and Fire Suppression

In addition to requiring primary and secondary electrical power, IT data centers rely on cooling for equipment along with ventilation, fire and smoke detection and suppression, cable management, physical security, work and storage space for spare equipment and parts, as well as physical buildings. Physical buildings need reinforced construction and a location above any flood zones; they also need to be tornado-proof and fireproof or at least fire resistant, Other considerations include isolation of different zones, insulation, possible use of waste heat to heat office space, natural light, building orientation and raised floors or overhead conveyances for power and cooling.

For larger data centers, separate rooms on the same or different floors separate IT equipment from mechanical and electrical devices. For example, servers may be in a room separate from disk and tape storage, with networking equipment spread among different rooms. Electrical switching, UPS battery backups, and related equipment including standby generators should be isolated in other areas, with power distribution units being near IT equipment.

6.4.1   Cooling and HVAC

With cooling and ventilation accounting for about half of all electrical power consumed in many data centers, any improvement in air flow and cooling will have a positive impact. For example, if 500 kW are being used to cool a data center, and raising the room temperature by 5°F can reduce energy consumption by 5%, that results in a savings of 25 kW, which might be enough to support near term growth or transition to more energy-efficient servers and storage. In general, 1 Btu is generated from 0.293 watt of energy; on average, 3.516 kW of energy requires 1 ton of cooling, or 12,000 Btu require 1 ton of cooling.

There are many different approaches and strategies for deploying CRAC and HVAC equipment, including around the edges of a room, at the edges and in the middle of a room, and in lines, with equipment arranged in hot and cold aisles. Another approach for areas that are short on floor space is to install cooling equipment on top of equipment racks or cabinets. Cooling can also be performed inside closed cabinets using forced air or liquid cooling. Top-of-rack cooling modules mount vertically above equipment racks and draw hot air from inside cabinets or from a hot aisle, discharging into a cool aisle to save space. Note that unmanaged or forgotten floor openings should be attended to in order to prevent loss of air flow.

Figure 6.4 shows a data center with an exposed ceiling and no overhead return air ducting for heat removal. In this example, hot air is pulled back in from CRAC units that produce cool air forced under the raised floor. The cool air under pressure exhausts through perforated floor tiles and is pulled into equipment cabinets arranged in hot and cold aisles. The cold aisle is the cooling row, where equipment air intakes face each other in a row, pulling cool air through the perforated floor tiles and exhausting hot air into the hot aisle. The hot aisles have the air exhaust side of equipment facing each other, with warm air being pulled to the ceiling and back to the CRAC units for cooling. Also under the raised floor are cable management and cable conveyance systems for networking and electrical power. Working with technology suppliers and consultants, air flow and temperature analysis can be performed to assess areas of improvement to optimize cooling. For example, hot spots, cold spots, as well as areas of air flow contention can be identified so that changes can be made. Changes can include raising or lowering temperature in certain areas, and rearranging equipment locations or exhaust patterns.

Cable management and cable conveyance solutions keep cables grouped together to simplify maintenance as well as to maintain good air flow. By keeping cables grouped together, under-the-floor blockage of air flow is minimized, enabling CRAC and HVAC systems to operate more efficiently. The same holds true for cabling within cabinets and equipment racks, in that organized cabling that does not impede air flow allows for more efficient cooling. Many new data centers are being built or remodeled with several feet of clearance under the raised floor, sometimes as much as 5

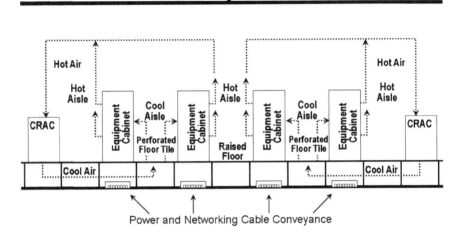

Figure 6.4   IT Data Center Cooling Without Plenums or Exhaust Ducting

to 6 feet or enough for some people to walk under the floors. A higher raised floor facilitates maintenance, growth, or addition of new cabling and better air flow.

A variation involves a suspended ceiling in rooms that house IT equipment. In the example shown in Figure 6.5, the suspended ceiling has perforated ceiling tiles with exhaust ducting to transfer heat back to CRAC units via plenums. The return air plenums direct warm air back to the CRAC units and avoid overcooling or overheating different zones in the room. As an alternative, exhaust air plenums that protrude down from the ceiling can be installed to improve air flow.

Another approach, shown in Figure 6.6, allows warm air to rise to a slopped or angled ceiling. Because hot air rises and cool air falls, the warm exhaust air rises to the ceiling and then, as it cools, is pulled down along the ceiling via large intake fans for cooling. Intake air is then forced through CRAC units or cooling coils similar to a radiator on an automobile, with cold air being forced under the raised floor. The cool air is then forced into the room via perforated floor tiles and into equipment in cold aisles, where the air is then exhausted and pulled upwards in a repeating process. With this open-air approach to cooling and ventilation, the air temperature will be lower under the raised floor than at floor or equipment height. Temperatures will be cooler in cold aisles and warmer in hot aisles. This type of

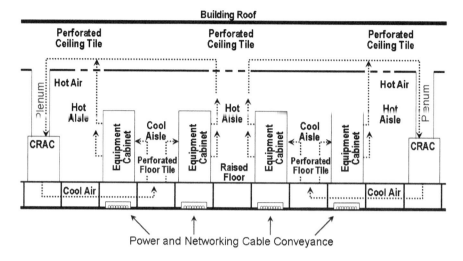

Figure 6.5   Overhead Ceiling with Air Return via Plenums to CRAC Units

approach is being used to supplement and enhance HVAC in large open facilities such as airport terminals in Hong Kong and Osaka, Japan.

A hybrid solution is shown in Figure 6.7, combining solar panels for supplemental power with a common cooling facility. The common cooling facility might be based on a heat pump exchanging heat via air, water in a drainage pond or adjacent river, or via underground cooling. For example, if enough open adjacent land exists, a large underground network of heat-

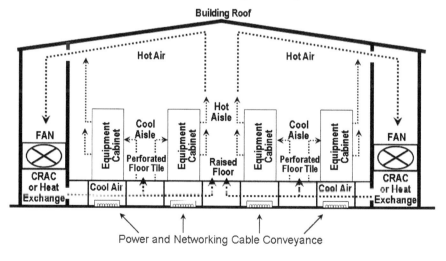

Figure 6.6   Cooling Using Sloped Ceiling, with Warmer Air near the Ceiling

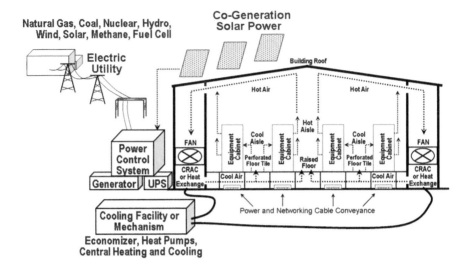

Figure 6.7   Hybrid Data Center with Alternative Cooling and Energy Sources

exchange piping can be installed deep enough to utilize the average below-ground temperature of around 55°F. Naturally, the depth will vary depending on the location as will the surrounding ground composition. In a rocky or hilly area, underground heat exchangers may not be practical.

Cooling systems can also be supplemented by using outside air for direct or indirect cooling via filters where practical. For example, indirect cooling would be useful for a facility such as the one shown in Figure 6.7, which is a closed-loop system with outside ambient air used for cooling water or coolant in the cooling system. Waste heat from equipment can be used to warm office or adjacent warehouse spaces during cool seasons. Natural light can be used for office and work spaces to reduce the need for electric lights during peak daytime periods. In very cold climates, waste heat can also be used for preheating outside air used by economizers for cooling.

Heat pumps can be configured for either heating or cooling. Heat pumps, in addition to providing a means of cooling or transferring waste heat, can also be used for dehumidification. Heat pumps can include geo-thermal leveraging of underground or other water source as a heat transfer technique, given proximity to an adequately sized, properly located water source. For extreme climate changes, ground and water sourced heat pumps can be used in place of air sourced heat pumps while tying the surrounding landscape into the heating and cooling ecosystem.

Another technique that is gaining popularity is precision cooling including open and closed systems using water or other liquid refrigerants. The idea in precision cooling is to cool as close to the heat source as possible to maximize cooling efficiency. By maximizing cooling efficiency, more heat can be removed using the same or less energy. For example, open cooling systems provide heat removable on a broader basis across multiple cabinets or racks in a cost-effective manor. However, open cooling systems are coarse in that large amounts of air are required to cool the heat source. Closed cooling systems can be implemented inside a cabinet, directing cooling close to where it is needed. The caveat is that closed cooling is more expensive and is tied to specific vendor implementation and support.

A new technique that is a variation on traditional water-cooled heat removal uses phased cooling. Phased cooling addresses issues with previous water-based cooling systems and potential coolant leaks in plumbing connectors and the subsequent risk of exposure to electrical equipment. Low-pressure coolants enable simpler flexible plumbing compared to traditional higher-pressure systems and also use less energy to move the coolant. By using coolants that converts to gas state when exposed to room temperatures, potential leaks are further reduced. The downside is that phased cooling systems require new plumbing. However, the upside is that phased cooling can be used more safely to support CRAC systems. As with other liquid-based cooling systems, heat removal from the water can be accomplished using traditional heat exchangers or external air economizers where the climate permits.

6.4.2   Physical Security

Security for data centers includes physical barriers with biometric and card scan access, motion detectors, auto-closing doors, and closed-circuit TV for security surveillance monitoring. Other security measures include secured shipping and receiving areas, separate staging and locked vendor spare parts bins or cages, as well as restricted access to the facilities in general. Physical structural considerations include being above ground and resilient to applicable threats such as earthquakes, hurricanes, and tornado force winds, along with protection from flooding. For example, in tornado-, hurricane-, or cyclone-prone areas, roofing materials should be heavy enough to withstand potential wind and air-pressure forces. Roofing shape is another consideration—for example, being able to drain off water or heavy snow. Fireproofing, detection, and suppression capabilities are another consideration, along with training or familiarization of local emergency response crews. Building alignment using natural surroundings can help reduce heating and cooling needs.

Structured cabling includes conveyance systems under raised floors or overhead systems for networking and power cables. Manual patch panels for cabling along with automated or intelligent physical media path panels help with additions, moves, or other reconfiguration changes involving cabling. Fan-out cables and wiring harnesses for high-density servers, networking, and storage systems help to reduce clutter associated with cabling and improve air flow. Cable tagging and labeling also helps in asset tracking and enabling infrastructure resource management activities.

Another suggestion is to isolate various types of equipment in different zones. For example, put backup magnetic tapes or disk-based backup devices in a separate room that is isolated from production systems, as a safeguard in the event of fire or other incident. An even better location for such equipment is off-site, or at least in another building, to meet compliance and data retention requirements.

6.4.3   Smoke and Fire Detection and Suppression

Smoke and fire are threats to data center equipment and the people who work in a facility. The fire triangle consists of oxygen, heat, and fuel—all need to be present for a fire. Remove the fuel, heat, or oxygen and the fire collapses. Timely detection and local suppression of smoking or smoldering items in a data center are important to safeguard people and equipment. Part of eliminating or avoiding smoke and fire problems is to identify potential causes, including overheating equipment, faulty electrical wiring, surge or grounding issues, overloaded circuits, short circuits, undervoltage, or wear and tear on cabling. Consequently, good cooling and heat-removal systems along with cable conveyance and management systems function to help protect against smoke and fire damage. Electrical fires are second to accidental fires started by smoking, discarded matches, and burning or hot embers. It is important to keep work and equipment areas clean and free of combustibles and away from ignition sources.

Minimizing the risk of smoke and fire should be a top priority in a data center. Automatic detection and notification as well as automated suppression are also vital aspects of fire protection. Advanced warning of rising or abnormal temperatures, abnormal air flow, and increased cooling activity combined with above-floor and below-floor sensors can alert personal to take proactive measures. By taking proactive measures, incidents can be contained and isolated without affecting other equipment and services being delivered in the same room or zone. Another benefit of early proactive measures, including automatically shutting down equipment, is to avoid or minimize equipment damage.

Early detection is important to address any smoldering problems as close to the source and as soon as possible. By applying topical or highly localized suppression or elimination of smoldering, smoking, or actually burning materials, damage to surrounding equipment can be minimized or avoided. For example, an air sampling solution may respond too slowly and then trigger a more widespread response, such as a high-volume dispersion of suppression agents that could damage equipment or backup media. With early detection, hand-held fire extinguishers with NAFS3 suppressant or other equipment using safe and environmentally friendly agents can be used to extinguish the fire or smoldering area without collateral damage to other equipment.

Smoke and fire detectors most often work by optical detection (photo-electrically), physical measuring using ionization, or by air sampling such as in VESDA detection systems. Detectors can also monitor carbon monoxide, natural gas, liquefied propane, or other explosive and flammable gases. Ionization detectors can detect small particles of smoke that are not visible, but they contain a small amount of radioactive americium-241 and therefore may not be preferable in some environments because of environmental health and safety concerns. Photoelectric devices, which are more environmentally friendly, may not be as effective for early detection compared to newer VESDSA solutions. VESDSA systems work by sampling air drawn into very sensitive sampler sensors above or below the floor to provide early detection. Given their ability to detect early signs of fire or smoldering material, VESDSA solutions may qualify for insurance incentives to help defray their cost and provide an economic stimulus for adoption and fire prevention.

Passive fire protection includes the installation of fire walls around the data center or around an individual room or zone, to prevent fire spreading to other areas. Special-purpose IT equipment can tolerate heat and subsequent water damage for certain periods of time and are used in remote or distributed environments or ultrasensitive applications. For example, fire- and waterproof or -resistant servers and disk drives can be used for sensitive information in harsh and risk-prone environments. Other possibilities include fireproof vaults and rooms for storing backup and archive data on either disk or magnetic tape. By separating backup, business continuity, or disaster recovery data in a separate room or facility, any possible smoke or fire suppression residue can be isolated from primary and backup data.

Beyond fire and smoke detection and notification systems are fire suppression systems. The main approaches for fire suppression are wet and dry pipe sprinkler systems and chemical systems. As the name implies, a wet system is a water- or fluid-based suppression system utilizing overhead sprinklers. Dry systems also use water or fluid, but the water is kept out of the pipes, using air pressure, until the system "trips," allowing water into the pipes to supply the sprinkler heads. Dry systems are used primarily in cool or unheated spaces that run the risk of freezing water-filled pipes. Chemical systems include the older halon-based systems and the newer FM200 type. Both wet and dry systems utilize water to suppress fires, but water does not mix well with the electrical systems found in IT data centers. Water-based systems should be used as backup suppression systems and not as a primary line of defense.

Chemical systems historically have used halons (organic halide compounds) as a means of suppressing fires; however, halons have been banned since 1987 as part of the Montreal Protocol prohibiting the use of certain substances that deplete the Earth's ozone layer. In addition to being bad for the environment, halons are also very dangerous for humans, depriving them of oxygen when discharged. In addition to addressing ozone depletion, a green activity in itself, the shift to environmentally safer fire suppressants also has ramifications for how physical facility and IT equipment are protected.

The Montreal Protocol does not mean that halon-based systems no longer exist; grandfathered systems are in use in numerous locations, but owners are being encouraged to eliminate them. However, no new systems are being installed, and existing systems that need recharging or replenishment must be done so with existing halon stocks. In place of halons, newer and safer agents such as Inergin, FM200, and NAFS3 agents are being used. These also leave fewer residues on equipment when used intentionally or accidentally. Chemical systems make for a good second line of defense for fire suppression, with the first line being prevention, early detection, and isolation on a localized basis. The National Fire Protection Association

Table 6.2   Classes of Fires and Applicable Suppressants

Symbol	U.S.	Europe & Australia	Fuel or Heat Source	Suppression	Application
	Class A	Class A	Ordinary combustibles	Remove heat, oxygen, or fuel, apply water to cool; remove oxygen with CO_2, use nitrogen or foam from extinguishers, or eliminate fuel.	Wood, paper, and other common combustible materials
	Class D	Class D	Combustible metals	Water can enhance metal-based fires; instead, suppress with dry agents to smother and remove heat.	Flammable metals such as titanium, magnesium, and lithium
	Class C	Class E	Electrical equipment	Avoid water or other conductive foam agents that put fire fighters in danger of electrical shock if electricity is not disabled. CO_2 and dry chemical agent suppressants should be used.	Electrical fires using nonconductive agents
	Class B	Class B	Flammable liquids	Avoid water, which causes fuel to spread. Inhibit chemical chain reaction with dry chemicals or suppression agents or smother with foam and CO_2.	Fires involving flammable liquids including grease, gasoline, and oil
	Class B	Class C	Flammable gases		Similar to above, focused on gases

(www.nfpa.org) is a good source for additional information about fire suppression and protection in general.

Table 6.2 shows the various classes of fires, with approaches for suppressing them.

Another consideration for fire detection and suppression, as well as general facility-wide site planning, is to protect against BLEVE events. A BLEVE event is a Boiling-Liquid/Expanding-Vapor Explosion, such as a leaking liquefied propane (LP) gas tank storing pressurized liquid that expands into a gas and explodes when exposed to a spark. Besides LP gas, other substances that when cooled are stored in a liquid state susceptible to BLEVE events include liquid nitrogen, refrigerants, and liquid helium.

As these and other substances are commonly transported by truck or railroad and are often stored locally, their presence should be considered when looking at locations for new data centers. Another consideration is that although an explosion, fire, or other severe event may not actually occur, officials will often close off an area in the event of a hazardous material spill or a leak of toxic, explosive, corrosive, or other dangerous substances, thereby preventing access to data center facilities. Thus, it's a good idea to try to locate a new data center away from highways, railroad tracks, pipelines, or other facilities where hazardous, corrosive, or explosive materials may be found.

General fire and smoke prevention, detection, and suppression topics include:

- Metallic and other water-based residue can damage equipment after it dries.
- Halon-based systems are costly to maintain and recharge, and they are discouraged.
- FM200 and NAFS3 solutions are being used instead of halon systems.
- CO_2 solutions cause thermal shock from rapid temperature drop, damaging electronics
- Halon, FM200, and other agents remove CO_2 from the fire triangle.
- There are concerns about FM200 with regard to adherence to the Montreal Protocol.

- Inergin is emerging as both an environment- and equipment-friendly solution.

- Heat and smoke sensors and alarms, including audible, visual, and electronic notification

- Isolation of different types of IT equipment into physically separate zones or rooms

- Good fire prevention habits—keep areas clean and test emergency procedures

6.4.4   Cabinets and Equipment Racks

IT equipment racks and cabinets utilize a standard height measurement of rack units (U's), where 1U = 1.75 inches or 44.45 mm. For example, a 42U cabinet is 73.5 inches tall. Racks and cabinets come in a variety of heights and depths, with a standard width of 19 inches. Cabinets can be open and free-standing for networking equipment or enclosed with locking doors, integrated power, and cooling including optional UPS for standby power, among many other options.

A common problem in many environments is lack of available floor space for new equipment, while existing equipment cabinets may not be fully utilized because of lack of available power, cooling, or standby power. To help boost available power in a given footprint, energy-efficient cooling, including liquid cooling technologies, can be used, thereby freeing up electricity to power equipment. To address internal data center polices limiting the amount of power or cooling per physical floor space footprint, new packaging and cabinet approaches are appearing on the market. Cabinets with built-in liquid or air cooling can enable denser equipment deployments per physical footprint, assuming adequate power is available. Side or in-line cooling shared between racks or dedicated to a rack can be used to boost available cooling per footprint. For example, current racks and cooling that are limited to 10 kWh could be boosted to 30 or even 60 kWh in the same or smaller footprint to support dense servers, storage, and networking technology.

Other considerations include:

- Weight per given physical footprint needs to be supported for dense equipment

- What maintenance tools are required for working with dense equipment in tall cabinets

- Do equipment cabinets need to be secured to flooring for earthquake or other threats

- Hot-swap maintenance capabilities of equipment and redundant components

- Existence of 80-plus efficient power supplies in equipment and cabinets

- Locking doors and access panels on cabinets to secure physical access to equipment

- Radiofrequency ID (RFID)-tagged equipment in cabinets to enable asset management

- Front, rear, and side access for racks, including pass-throughs for cables between cabinets

- Integration of equipment and cabinet alarms and sensors for monitoring and management

- Automated power down or enablement of power-savings modes for energy efficiency

- Blanking and filler plates for cabinet air flow management

- Cable management within cabinets to allow efficient air-flow and cooling

6.4.5   Environmental Health and Safety Management

As pointed out in Chapter 1, the green gap with regard to IT data centers also encompasses environmental health and safety (EHS), including e-waste and recycling. For data centers, in addition to reducing the amount of electrical power used or making more efficient use of available electrical power to support growth by reducing cooling demands, other actions include:

- Reducing the amount of water required for cooling

- Leveraging technology and digital media disposal services

- Eliminating hazardous waste material from discarded IT equipment

- Recycling paper from printers and copiers as well as equipment shipping cartons

- Removing equipment containing mercury, bromine, chlorine, and lead

- Implementing ISO 14001 and OHSAS 18001 EHS and related systems

- Adhering to other existing and emerging EHS and emissions-related legislation

- Measuring vendors or suppliers on their EHS and related management programs

6.5   Data Center Location

A common myth has all new data centers in the United States being built in the Columbia River Valley of Washington state. Reality has new data centers springing up all over the country, including in Ohio, Illinois, Iowa, Indiana, North and South Dakota, Wisconsin, and Minnesota, where a large number of generation facilities exist with large nearby coal fuel sources as well as generally unrestricted power transmission and networking bandwidth. Given the relative proximity to power-hungry and growing IT resources in California and surrounding areas, Las Vegas, with the adjacent Hoover Dam, is seeing more data centers being built, as are other areas in surrounding states.

The Midwest is not alone in seeing growth; Google has a high-profile site on the Columbia River, and Microsoft has a large site in Ireland as well as one outside Chicago. Power is an important consideration, but so too is the cost of land, isolation from relevant threat risks, and workforce and network capabilities.

Some things to look for in a data center location include:

- Access to adequate electrical power generation and transmission sources

- The cost of energy for primary and secondary power and cooling needs

- Network bandwidth adjacency along with redundant paths for high availability

- Real estate and facilities construction or remodeling costs

- Proximity to affordable and skilled workforce or housing for staff

- Safety from floods or other acts of nature
- Distance from applicable threat risks including acts of man or nature

6.6   Virtual Data Centers Today and Tomorrow

Virtual data centers can consolidate underutilized servers, storage, and networking equipment while also supporting scaling for growing applications. For large applications that require more resources than a single server, storage, or networking device, virtualization can be used as a management and technology abstraction layer. The abstraction layer enables routine planned and unplanned maintenance while minimizing or eliminating disruptions to applications and their users.

Virtualized abstraction and transparency also enable older technology to be replaced by newer, faster, higher-capacity, denser, energy-efficient technologies to maintain or recapture power, cooling, floor space, and environmental footprint capabilities. For large-scale or extreme megascale bulk storage or compute-intensive server farms or grids, large blocks of servers or storage can be replaced on an annual basis. For example, assuming that technology is kept for 5 years, each year one-fifth of the technology is replaced with newer, higher-capacity, faster, smaller-footprint, denser, and more energy-efficient technology.

For large-scale environments faced with rapid growth or already large server, storage, and network deployments, a new approach to bulk resource installation and implementation uses large shipping containers. Putting data centers on truck trailers is a practice that, over the past couple of decades, has been used by some vendors to provide "data centers on wheels" for disaster recovery needs. A growing trend is for IT resource manufacturers, particularly server vendors, to provide preconfigured, "prestuffed" data centers in a box using industry-standard 20-foot shipping containers (see Figure 6.8). Instead of using shipping containers to transport the cabinets and server, storage, and networking components, cabling, power, and cooling components to a customer site, shipping containers are preconfigured as ready-to-use mini-data centers or modular computer rooms or zones for a larger data center.

Preconfigured data center solutions have been in existence for several years, with servers, storage, networks and associated cabling, power, and

Figure 6.8   Data Center in a Shipping Container (Courtesy www.sun.com)

cooling already integrated at the factory. What differs among existing preintegrated solutions, sometimes called data centers or SANs in a can, is density and scale. Existing solutions are typically based on a cabinet or series of
cabinets preintegrated with all components prestuffed or ready for rapid
installation at a customer's site.

The new generations of data centers in a box are based on the shipping
industry standard intermodal container. Containers are called intermodal
because they can be transported with different modes of transportation,
including ship, rail, and truck, or by large cargo-carrying aircraft, without
having to unload and repack the cargo contents. Given the diverse use of
intermodal containers, standardization has resulted in robust and sturdy
containers commonly used for shipping high-value cargos. Interior dimensions of a standard 20-foot intermodal container are approximately 18 feet
10 inches by 7 feet 8 inches wide and about 7 feet 9 inches tall. These intermodal containers are commonly seen in 20-, 40-, 45-, 58-, and 53-foot
lengths on cargo container ships, railroad flat cars, or intermodal double
stack (two containers high) over-the-road tractor trailer trucks.

While the basic exterior shape and dimension are standard, as are
construction quality and strength, vendors' solutions vary, including electronic magnetic interference (EMI) along with radiofrequency interference (RFI) shielding, fireproofing, and other options. Similar to

refrigerated, freezer, or heated containers used for transporting perishable food or other items on ships, trucks, or trains, containerized data centers are designed to plug into data center power and cooling facilities. Some containers are configured with adequate power to support up to 30 kWh of power per rack for ultradense servers, assuming adequate power for the container itself is available.

Server vendors, including IBM, Sun, and Verarri, among others, are not the only ones leveraging these large shipping containers. Large consumers of IT resources such as hosting sites, managed service providers, and large-scale content providers such as Microsoft are using tailored vendor solutions or customized, built-to-specification container-based modules.

Benefits of using preconfigured large container-based data center modules include:

- The shipping container is an ultralarge, field-replaceable unit for technology upgrades.
- IT resources including servers, storage, and networking are preinstalled and integrated.
- Shorter installation and implementation time is required at the customer site.
- Rapid deployment of new or upgraded data center capacity is possible.
- Containers plug into facility power, cooling, networking, and monitoring interfaces.
- Power, cooling, and cabling are preintegrated in the large modular container.
- Overhead and other built-in conveyance and cabling schemes can be used.
- Volume deployment of server, storage, and networking equipment is supported.
- The size of receiving, shipping, and equipment staging areas can be reduced.
- Raised floors can be eliminated, using concrete slabs with overhead utility conveyance hookups instead.

6.7   Cloud Computing, Out-Sourced, and Managed Services

IT resources including servers, storage, networking, and applications that are hosted remotely in an abstracted and virtualized manner are commonly referred to as *cloud* or *managed services*. Although the implementations and marketing initiatives of cloud computing, storage, applications, and software as a service have evolved, they are in essence variations of previously available managed or hosted services. For example, instead of maintaining IT facilities, servers, storage, networks, and applications, organizations can opt to have business applications run remotely as a service on someone else's hosted virtual environment, accessed via a public or private network. The term *cloud* stems from the generic nomenclature that is often used to represent various services or functionalities. For example, in diagrams, a drawing of a cloud may be used to represent some infrastructure item such as a network, the Internet, or some other generic entity, without showing details. Similarly, cloud computing or cloud storage leverages this cloud or generic nomenclature to abstract the services that are actually being delivered. Cloud-based services, including cloud computing and cloud storage, are discussed in later chapters.

A variation of a virtual data center is to move applications and data to a cloud or virtual data center that is managed and provided as a service. Entire IT applications and data storage capabilities can be out-sourced, including being physically moved to a hosting or managed data center or managed on-site at a customer location. Another variation is leveraging core IT resources for sensate, mission-critical strategic applications and data while using managed services for business continuity and disaster recovery, archiving of inactive or compliance data, or for specialized services including Web portals, e-commerce, social networking, enterprise 2.0, and other Web 2.0-related applications.

Cloud and managed service provider names and acronyms may be new; however, the fundamental concepts go back most recently to the ISP, ASP, SSP, or generic xSP models of the Internet dot.com craze. Older variations include on-site managed services, time-sharing, and service bureaus, as well as traditional out-sourcing to a third-party firm. What has changed besides the names and buzzwords are the types of services, cost structure, network and processing performance improvements, and availability. What remains

constant is an increasing amount of data to move and store for longer periods of time along, with existing and emerging threat and security risks.

Leveraging cloud servers, hosting, and managed service providers enables
organizations to focus on core business functions or avoid costs associated
with establishing an IT data center. Some business models rely extensively on
out-sourced or managed services, including contract manufacturing for virtual factories, managed payroll and human resource functions from companies like ADP and Paychex, or email services from other providers.

While smaller businesses can off-load or defer costs to managed service
providers for functions including data protection, email, Web hosting, and
other functions, large organizations can also leverage online, managed, and
traditional out-sourcing services as part of an overall IT virtual data center
strategy. For example, legacy applications associated with aging hardware
can be shifted to a managed service provider whose economies of skill and
operating best practices may be more cost-effective than in-house capabilities. By shifting some work to third parties, internal IT resources including
hardware, facilities, and staff can be redeployed to support emerging applications and services. Similarly, new technologies and applications can be
quickly deployed using managed service providers while internal capabilities, staffing, and associated technologies are brought up to speed prior to
bringing applications back in-house.

Cloud and managed services can be thought of as service delivery capabilities or personalities of IT resources, such as servers for compute or running applications as well as data repositories for storing data. Some
examples of service personalities include email, digital archiving management systems, online file storage, portals for parking your photos and videos, backup, replication, and archiving storage targets, among others.

A benefit of cloud, managed, utility, or hosted services is to off-load IT
functions from organizations, enabling them to focus on their core business.
Essentially, costs and control are shifted to a third party instead of an organization establishing and managing IT resources on its own. If an IT data
center is optimized for productivity and efficient resource use, the cost to
use a third-party or external service may be higher than using internal
resources. On the other hand, internal resources can be supplemented by
external services for email, Web hosting, business continuity and disaster
recovery, or other specialized services. Third-party services can also be used
for surge or on-demand supplement resource capacity to shift certain applications while updating or restructuring existing data centers.

When considering out-sourcing, pricing, particularly surcharges for excessive use or changes to data, should be evaluated along with service-level agreements and redundancy and compliance. Fees are typically some amount per month per gigabyte plus fees for uploads and downloads, similar to how a Web hosting or managed service provider may charge. Additional fees may apply for different levels of service, including faster storage and network access, improved data resiliency and availability, or long-term retention. Access mechanisms can include Network File System (NFS) or Windows Common Internet File System (CIFS)-based NAS, Microsoft .NET, or Web server Hypertext Transport Protocol-HTTP, among others. Other optional fees can apply to extended technical support, domain name recognition and hosting, client billing and credit card processing, secure socket layer (more commonly known as SSL) and security encryption certificates, and special storage, or application services.

The prospective service provider's financial and business stability should also be considered, including how data can be retrieved should the service cease to exist. As with any emerging technology or solution provider, look for organizations that are flexible and that also have a growing list of active customers. It is important to look at the number of customers actually using the service actively, as opposed to just the marquee names. Banks such as Wells Fargo (vSafe) have announced or are already providing managed services including virtual digital safe deposit boxes for storing important documents. Other providers of digital repositories include telecom providers such as ATT. Cloud and managed service vendors as of this writing include Amazon, EDS, EMC, Google, HP, IBM, Iron Mountain, Microsoft and Seagate, among others.

Regarding providers of cloud and managed solutions, look into how they will ensure that different customers' data and applications are kept separate, not only on disk storage systems and file systems but also when in memory on virtual servers. Another thing to consider is the establishment of alternate copies of critical data elsewhere, independent of the copy entrusted to the cloud or managed solution provider, to ensure continued access as well as to guard against data loss or loss of accessibility to the data if a cloud service goes offline unexpectedly. Additional things to consider about cloud, out-sourced, managed service, and utility-based virtual data models include:

- Security and information privacy of stored data
- Encryption of data, rights access, and authentication to data and applications
- Web-based SSL security of data in transit

It is important to determine if moving applications or data to a cloud is simply moving a problem or is an opportunity to improve the overall business. It is also necessary to keep in mind that a cloud-based service needs to make a profit and needs to rely on its own host or managed service provider to house its service or leverage its own facility including equipment. Its goal is to keep costs as low as possible, spreading resources across multiple users or subscribers, which could lead to performance or resource contention issues. Performance and latency or delays in accessing information from a cloud or network-based service provider also need to be kept in perspective. Mission-critical or high-risk applications may not be well suited for a cloud or Web-based model, however, they can be safely hosted in managed facilities or outsourced data centers using secure best practices.

Cloud and virtual data centers still require physical resources, including servers, storage, and networks, as well as a facility to house the hardware, and software to deliver and manage the service. Similar to a manufacturing company performing a make-versus-buy analysis along with building in-house or using a third party considerations, IT organizations moving forward should analyze their needs and requirements for in-house or external cloud and managed provider services.

6.8 Data Center Tips and Actions

Short of building expensive new data center facilities with adequate generating capacity near energy sources and unrestricted transmission networks with potentially lower energy costs, IT organizations need to look at other near-term facilities options. Even if plans include a secondary (or tertiary) new data center now or in the future, available power and CRAC cooling capacity will need to be balanced with increased demand to support growth.

Data center assessments range from simple, quick, walk-through observations to more detailed air flow measurements and modeling with computational fluid dynamics software to optimize equipment placement, cool air flow delivery, and effective heat removable. More advanced techniques include increasing temperature set points to a balanced point between ambient temperatures that require less power and avoiding IT equipment that draws excessive power to keep components cool. However, exercise caution, conferring with appropriate engineers and manufacturers' support personal for temperature and cooling guidelines for specific equipment, to prevent accidental overheating and subsequent equipment damage. Other techniques include replacing unneeded vented and slotted raised floor tiles as well as removing old cables that block air flow, increasing the height of raised floors during data center remodeling, and building out to allow for growth and increased air flow.

Exercise caution when using nameplate metrics on IT equipment, including cabinets. Space in a cabinet or rack may not be fully utilized, resulting in extra cabinets or racks further tying up or wasting data center floor space. For example, the nameplate on a 4U server may indicate a breaker or circuit of 1 kVa, but actual maximum or surge power consumption is only 0.75 kVa. Carrying this simple example further, assuming a cabinet is capable with its internal power distribution units of supporting 6.5 kVa, using nameplate metrics would allow only six 4U servers consuming 6 kVa in 24U, or a little over half of the cabinet space. Using 0.75 kVa per 4U server would allow eight 4U servers in the same equipment rack or cabinet using 6 kV or power and 32U of the 40U cabinet. Granted, this is a simple example and best practices should include consultation with manufacturers' installation guides and services organizations along with facilities personal and electrical engineers before actual configuration. The point is that in a quest to boost resource utilization, look beyond simply consolidating unused servers with virtualization. Look for other opportunities to safely boost resource utilization of data center floor space, cabinets, and equipment without negatively impacting performance, availability, capacity, or environmental health and safety.

Investigate raising data center temperatures a few degrees, after conferring with technology suppliers for best practices, to reduce energy consumption. For example, depending on age and configuration of equipment, raising the ambient temperature in a large computer room a few degrees can result in a couple of percentage points decrease in electrical energy consumption. Depending on needs, saved or reclaimed power can be used to reduce operating expenses or to support growth and expansion. Note that some additional power may be used by cooling fans periodically, depending on ambient room temperature; however, depending on cooling savings, the increase in fan power draw may be negligible. When combined with variable power management, switching to lower-power modes will draw less power, resulting in less cooling and therefore more savings.

Put a technology refresh plan in place on a 36-, 48-, or 60-month cycle, depending on finances and principles of operation, to turn over older, less energy-efficient technologies to boost performance and capacity to sustain growth and fit into existing or future PCFE footprint constraints. Replace older equipment with newer, more energy-efficient and higher-performance technology to allow consolidation of underutilized technologies (servers, storage, and networks) or to support scaling of workloads that need to grow.

Data center facilities and associated technology vendors include Anixter, Ansul, APC, BMC, CA, Caterpillar, Cummins, Detroit, Eaton, Emerson/Liebert, Enigmatec, Firelock, GORE, Honeywell, HP, IBM, IOsafe, Kohler, Microsoft, Panduit, Powerware, Rackable, Rittal, Sanmia, Siemon, Spraycool, Summit Fire, Tripplite, Rackable, Verari, and Writeline among others.

6.9 Summary

Data centers continue to evolve, and, with a growing spotlight on green PCFE-related issues, there are opportunities to improve productivity. Data centers will continue to adapt to support changing business needs, boosting efficiency and productivity to remain competitive. Factories need to be reinvested in and retooled to remain competitive; so, too, must IT data centers be kept up to date to remain competitive and be a corporate asset rather than a cost center. This means that investments need to be made to improve the efficiency and use of resources, such as electrical power, so more work can be done with available resources.

Action and takeaway points from this chapter include:

- Existing environments can evolve to become green and virtual environments.
- Perform an assessment, including air flow and temperature analysis.
- Investigate where power is being used as well as possible upgrades to newer technology

- Consolidate where practical; implement newer, faster technology where needed.

- Investigate safely raising air temperature in some zones.

- Practice infrastructure resource management to maxize IT resource usage and availability.

- Apply data management to reduce the impact of an expanding data footprint.

- Leverage alternative energy sources and co-generation.

- Capitalize on energy incentives and rebate opportunities.

- Integrate technology and facilities planning across IT.

Chapter 7

Servers—Physical, Virtual, and Software

Software that truly does not require hardware will be revolutionary.

In this chapter you will learn:

- About the demand for more servers that consume more power and generate more heat
- The benefits and challenges associated with increased server densities
- Differences between physical and virtual servers
- The many facets of server virtualization
- Differences between clustering and grid computing

The importance of this chapter is understanding the relationship between the need for more processing capability and the consequent energy demands versus the need to improve energy efficiency to reduce power and cooling costs and their environmental impacts.

Computers—also known as blade centers or blade servers, desktops, laptops, mainframes, personal computers (PCs), processors, servers, and workstations—are a key component and resource of any data center or information factory. Computers run or execute the program software that performs various application and business functions. Hardware does not function without some form of software—microcode, firmware, operating or application software—and software does not function without hardware. The next truly revolutionary technology, in my opinion, will be software that does not require hardware and hardware that does not require software.

7.1   Server Issues and Challenges

In the typical data center, computers are the second largest consumer of electrical power, after cooling. In addition to requiring power, cooling, and floor space, computers have an environmental health and safety footprint in the form of electronic circuit boards, battery-backed internal power supplies, and other potentially hazardous substances that need to be recycled and disposed of properly. The larger the computer, the more components it will have; smaller computers, such as laptop or desktop workstations, have fewer and smaller components, although they do have batteries and monitors or screens.

In general, servers need to:

- Be available and able to run applications when needed to meet service requirements
- Fit into a smaller physical, power, and cooling footprint—doing more work efficiently
- Achieve more performance per watt of energy
- Support applications that continue to grow and add new functionality and features
- Be configured and redeployed to meet changing market and business dynamics
- Be seen as a key resource to enable IT services effectively

In 1965, Gordon Moore, who co-founded chip giant Intel, made an observation, now known as *Moore's law*, that the number of transistors per square inch on an integrated circuit (IC) had doubled every year since the IC (or chip) was invented. Moore's law predicted that the trend would continue into the forseeable future, which has, for the most part, held true over the past 40 years later (Figure 7.1). Although the pace has slowed a bit for processors, the amount of data density doubling every 18 to 24 months is also included under Moore's law. What is important to understand about Moore's law is that the general industry consensus among IT professionals, manufactures, analysts, and researchers is that the current trends will continue to hold true for at least a few more decades. Thus, future processor and storage needs can be estimated by looking at past supply and demand.

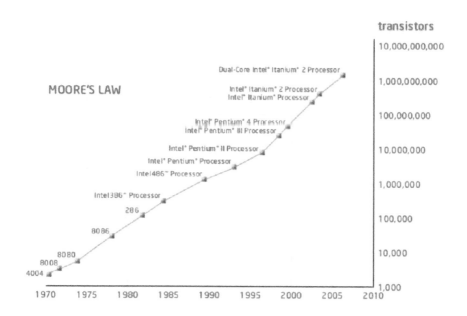

Figure 7.1   Moore's Law (courtesy of Intel)

Servers are a key resource for supporting IT-based services. Some drivers that are pushing the continued demand and limits of available processing performance include:

- Expansion and enhancement of existing applications with faster response time

- Higher volume of transactions, messages, downloads, queries, and searches

- Marketing analysis and data mining to support targeted marketing efforts

- Simulation and modeling for product design and other predictive services

- On-demand media and entertainment services, including gaming and social networking

- Service-oriented applications to support data mobility, Enterprise, and Web 2.0

- Expanding medical and life science applications, along with essential emergency support services

- High-performance scientific, energy, entertainment, manufacturing, and financial services
- Increasing defense- and security-related applications
- Massive scale-out and extreme computing—the opposite of consolidation

Some common issues and challenges of servers include:

- Continued sprawl of servers in environments with reduced power, cooling, floor space, and environmental (PCFE) capabilities
- Multiple low-cost or volume servers dedicated to specific functions or applications
- Underutilized servers and increasing software and management costs
- Servers that cannot be consolidated because of application, business, or other concerns
- Applications that have or will outgrow the capabilities of a single server
- Performance, availability, capacity, and energy consumption
- Timely and complex technology migration with associated application downtime
- Support for new computing and IT service capacity and models
- Growing awareness of the importance of data and information security concerns
- Dependence of IT services users on the availability of applications when they are needed
- The shifting focus from disaster recovery to business continuance and disaster avoidance

Applications or computer programs need more computing power to perform more complex tasks, as seen in Figure 7.2. For example, a transaction or event requires more code to be executed to perform a particular function, and more functions need to be performed on different interde-

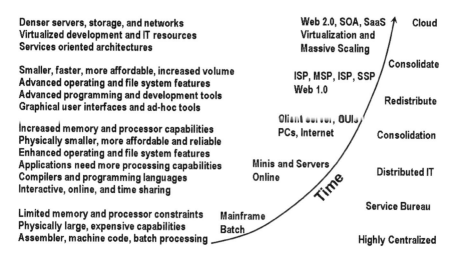

Figure 7.2   Growth and Demand Drivers for Server Compute Capabilities

pendent systems, particularly with complex applications that include rich media, validation, and rules.

Figure 7.2 shows how computing capabilities have evolved over time from standalone mainframes supporting batch processing and requiring timely low-level programming to highly optimize applications that maximized limited resources. Moving from lower left to upper right in the figure, the continuum shows how the advent of minicomputers and PCs with increased memory, processing, and input/output (I/O) capabilities enabled broader adoption of computers along with higher-level programming languages and development environments. Also enabled were easier-to-use graphical user interfaces (GUIs), boosting productivity but also requiring more computing power and memory capacity.

Note that server computing capability is often referred to as compute or processing power. While there is a correlation to the amount of power drawn by a processor to achieve a given level of compute operations or cycles, care should be taken to keep in the proper context compute processing capacity or compute power versus the amount of power required and used by a server.

Another item to factor in is that many applications are now developed and written in frameworks and nth-generation languages or development suites to provide a level of abstraction that can speed and aid development and alignment of business rules. For example, consider how the code to perform the same transaction has changed over the years, from assembler or macro code to FORTRAN to COBOL or Basic to C language or other tools and environments including Perl, Ruby on Rails, and others.

One of the by-products of application development frameworks has been to speed up the development of more complex systems, often leveraging lower-cost desktop or small servers rather than tying up resources on large and expensive servers or mainframes. Each successive generation of tools and software development frameworks, as shown in Figure 7.2, has enabled faster development and increased productivity for application developers. At the same time, however, more lines of code are being generated to support a given function, transaction, or event, requiring more processing power to support the behind-the-scenes generation of software code along with applicable runtime and executable libraries.

The result of more processing cycles being required to perform a given function is a productivity gap. This gap is being addressed by using more servers and more computing capabilities, which consumes more electricity and generates more heat.

Servers and computers in general generate heat by using electricity:

- Increased density of chip-level logic gates and corresponding switching speed leads to generation of more heat.

- Processing chips and support chips can consume 30% or more power per server.

- Power supplies and cooling fans can account for 30–40% of the power used by a server.

- Video monitors for laptops and workstations may require large amounts of power.

- Power used for memory for storing data and programs is almost continually increasing.

- Support chips for memory, I/O, and other peripheral access and internal busses consume power.

- Adapter cards for networking and I/O, including transceiver optics, consume power.

- Power supplies, power distribution, and cooling fans must be powered.

- Optional battery backup **uninterruptible power supplies (UPS)** may be part of a server.

Servers consume different amounts of electrical power and generate various amounts of heat depending on their operating mode, for example, zero power consumption when powered off but high power consumption during startup. Servers use different amounts of energy when running with active or busy workload and less power used during low-power, sleep, or standby modes. With the need for faster processors to do more work in less time, there is a corresponding effort by manufacturers to enable processing or computer chips to do more work per watt of energy as well as reduce the overall amount of energy consumed.

To address the various server demands and issues, techniques and approaches include:

- Consolidation of underutilized servers using virtualization software
- Shift from consolidation to using virtualization for management transparency
  - Enable rapid provisioning or reprovisioning of servers to meet changing needs
  - Facilitate business continuity and disaster recovery testing and implementation using fewer physical resources
  - Remove complexity and enable faster technology upgrades and replacement
  - Reduce application downtime for scheduled and unplanned maintenance
- Awareness and demand for clustered, grid, and cloud services-oriented applications
- Desktop virtualization for consolidation of thin client applications and dynamics
- Technology to address PCFE issues, including 80%-plus efficient power supplies
- Improved best practices and rules of thumb to boost efficient use of energy
- Energy Star compliance for servers as a means to compare server energy efficiency
- Recycling of retired equipment as well as removal of hazardous substances and consideration of environmental health and safety.

In addition to boosting energy efficiency by doing more work per watt of energy consumed, computer chips also support various energy-saving modes such as the ability to slow down and use less energy when there is less work to be done. Other approaches for reducing energy consumption include adaptive power management, intelligent power management, adaptive voltage scaling, and dynamic bandwidth switching, techniques that are focused on varying the amount of energy used by varying the performance level. As a generic example, a server's processor chip might require 1.4 volts at 3.6 GHz for high performance, 1.3 volts at 3.2 GHz for medium performance, or 1.2 volts at 2.8 GHz for low performance and energy savings. Intel SpeedStep® technology is an example that is commonly used in portable computers and mobile units, as well as some desktop workstations and servers, to reduce energy when higher performance is not needed.

Another approach being researched and deployed to address PCFE issues at the server chassis, board, or chip level includes improved direct and indirect cooling. Traditionally, air has been used for cooling and removing heat, with fans installed inside server chassis, cabinets, and power supplies or even mounted directly on high-energy-consuming and high-heat-generating components. Liquid cooling has also been used, particularly for large mainframes and supercomputers. With the advent of improved cooling fans, moderately powered servers and relatively efficient ambient air cooling, liquid-based cooling was not often considered for servers until recently.

Given the continued improvements in density, putting more processing power in a given footprint, more heat needs to be removed to keep component temperatures within safe limits. Consequently, liquid cooling is reappearing in many forms for cooling whole data centers as well as individual cabinets, servers, and even individual chips. Liquids such as water can cool approximately seven times more efficiently than air-based cooling. Some challenges of liquid-based cooling are that plumbing needs to be introduced to the system, adding complexity and the risk of a leak. Keeping in mind that fluids and electricity do not mix, concerns about packaging and plumbing have resulted in a general preference away from liquid cooling. However, with the higher densities and heat being generated today, new approaches to liquid-based cooling are being considered.

One of these new approaches is liquid pumps. With regular liquid-based cooling, the coolant is chilled and then put in contact with the surface or air to be cooled, raising the coolant temperature. The coolant is re-circulated, re-cooled and the process continues. Energy is required to cool the coolant as well as energy needed to push the coolant through the plumbing network. With pump-based cooling solutions, liquid coolant is moved through flexible piping at low pressure to come in contact with cooling surfaces, such as heat sinks or other cooling devices, to remove heat. Unlike traditional water liquid cooling systems under higher pressure, low pressure pumped coolants convert to gaseous state when at room temperature to avoid liquid leaks and spills. The result is that more flexible plumbing or distribution technologies can be used, including the ability to cool at the board and chip level without putting liquids in contact with sensitive technologies. A benefit of pump-based liquid cooling is that heat can be removed even more efficiently than traditional liquid based cooling and even more effectively than air-based cooling.

Liquid cooling at the chip level may involve attaching on-chip heat sinks to cooling systems, layering ultrathin pumped cooling using liquid or gas coolant between silicon chip layers, and other approaches. The technology is still fairly new but will probably be more widely used because the most efficient way to cool is at the heat source itself.

With each successive generation of new servers and processors, more processing capability or computing power has become available in a smaller footprint. As shown in Figure 7.3, a smaller footprint means that less power per cycle or operation is used for a physically smaller footprint. However,

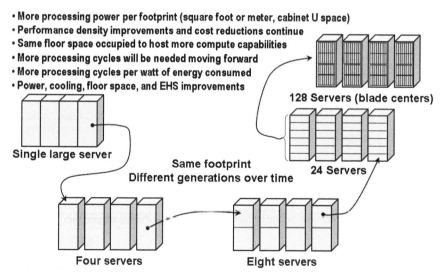

Figure 7.3   Server Footprint Evolution—Doing More in the Same or Smaller Footprint

while more processing power can be put within a given physical footprint, the aggregate power and cooling demands on a given footprint continue to rise to meet the demand for more computing capability.

In some environments, the result of reducing the size while increasing performance and lowering costs for servers can mean more available floor space or energy savings. On the other hand, in environments that are growing and adding more demanding applications and supporting larger amounts of data to be processed, any savings in power, cooling, or floor space is used for expansion to support and sustain business growth. For example, as in Figure 7.3, reducing physical size and increasing processing performance will result in lower operating costs and recovered floor space for some organizations. However, for growing businesses, that floor space will be filled by expansion of denser technologies that need more electrical power and cooling than in the past. Consequently, as servers become faster, there will be an emphasis on further reducing the electrical power consumed and the heat that needs to be removed while enabling more work to be done.

7.2   Fundamentals of Physical Servers

Servers vary in physical size, cost, performance, availability, capacity, and energy consumption, depending on their specific features for different target markets or applications. Packaging also varies across different types of servers, ranging from small hand-held portable digital assistants or PDAs to large-frame or full-cabinet-sized mainframe servers. As shown in Figure 7.4, servers, regardless of actual implementation architecture and specific vendor

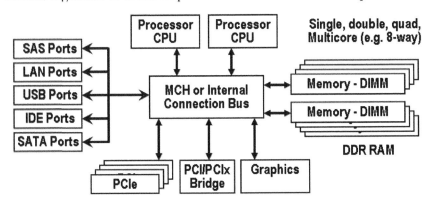

Figure 7.4   Generalized Computer Hardware Architecture

nomenclature, generally have a processor, memory, internal busses or communication chips, and I/O ports for communicating with the outside world via networks or storage devices.

The specific components and number of components vary according to the type of server, but all servers usually have one of more of the following:

- One or more **central processing units (CPUs),** also known as processors
- Some amount of main **random-access memory (RAM)** or **read-only memory (ROM)**
- Internal communication busses and circuitry
- **Basic Input/Output System (BIOS)** and/or a management console subsystem
- Attachment points for keyboards, video, and monitors
- I/O connectivity for attaching peripherals such as networks and storage
- Power supplies and cooling fans

Additional possible components include:

- Video monitors and displays
- **FLASH-based solid-state disk (SSD)** devices or adapters
- Internal or external disk storage, such as **hard disk drives (HDDs)**
- **Removable hard disk drives (RHDDs),** floppy disk or tape devices
- Keyboards, mice, or other pointing and user interaction devices
- Specialized processors or accelerators including graphics cards
- I/O **host bus adapter (HBA)** and **networking interface cards (NICs)**
- Integrated **redundant array of inexpensive (or independent) disks (RAID)** on the mother board, known as **ROMB,** or other enhanced functionality
- Mirrored or RAID-protected main memory for high availability

- Built-in wireless networking communications such as WiFi, WiMax, and 3G
- Specialized cooling and heat-removing capabilities
- Battery-backed UPS or DC power capabilities

Computers or servers are targeted for different markets, such as small office/home office. small and medium business, large enterprise, and ultralarge-scale or extreme scale, including high-performance supercomputing. Servers are also positioned for different price bands and deployment scenarios.

General categories of servers and computers include:

- Laptops, desktops, and workstations
- Small floor-standing towers or rack-mounted 1U and 2U servers
- Medium-size floor-standing towers or larger rack-mounted servers
- Blade centers that house blade servers, blade networks, and I/O switches
- Large-size floor-standing servers, including mainframes
- Specialized fault-tolerant, rugged, and embedded-processing or real-time servers

Depending on their use, servers may have many different names: email server, database server, application server, Web server, video server, file server, network server, security server, backup server, storage server and so on. What defines the type of server is the software used to deliver a particular service. Sometimes the term *appliance* is applied to a server, indicating the type of service the combined hardware and software solution are providing. For example, the same physical server running different software could be a general-purpose applications server, a database server running Oracle, IBM, or Microsoft databases, an email server, or a storage server.

This terminology can lead to confusion when looking at servers in that a server may be able to support different types of workloads and thus might be considered a server, storage, networking, or application platform. Which type of server this one is depends on the type of software being used on it. If, for example, storage software in the form of a clustered and parallel file system is installed to create highly scalable **network-attached storage (NAS)** or cloud-based storage service, then the server is a storage server. If the server has a general-purpose operating system such as Microsoft Windows, Linux, or UNIX and a database on it, it is a database server.

Although not technically a type of server, some manufacturers use the term *tin-wrapped software* to avoid classification as an appliance, server, or hardware vendor; they want their software to be positioned as turnkey solutions, not software-only solutions that require integration with hardware. The idea is to use off-the-shelf, commercially available general-purpose servers with the vendor's software technology preintegrated and installed. Thus, tin-wrapped software is a turnkey software solution with some "tin," or hardware, wrapped around it.

A variation of the tin-wrapped software model is the software-wrapped appliance or a virtual appliance. In this case, vendors use a virtual machine (VM) to host their software on a physical server or appliance that is also being used for other functions. For example, database vendors or virtual tape library software vendors might install their solution into separate VMs on a physical server with applications running in other VMs or partitions. This approach can be used to consolidate underutilized servers, but caution should be exercised to avoid overconsolidation and oversubscription of available physical hardware resources, particularly for time-sensitive applications. *Oversubscription* is a means by which fewer physical resources are needed to support a client base if, on average, only a certain number of users of a service are active at a time. Oversubscription can reduce costs at a given level of performance, but if all clients become active at the same time, then oversubscription occurs. Obvious examples are cell phone networks and the Internet, both of which are designed to handle an average number of concurrent users. If many more users become active at the same time, busy signals or long delays can occur.

7.2.1   Central Processing Units

There are many different types of computer processor chips, varying in performance (speed), energy consumption for mobile or other uses, physical size, extra features for virtualization or security, I/O optimization, number of cores (single, dual, quad, or eight-way) or processors in a single chip, amount of cache and on-board memory, along with energy management features in hardware, via firmware or external software such as operating systems or layered applications.

Basic computer processor architecture includes the **instruction set architecture (ISA)** with **program status words (PSWs),** a **program counter (PC),** and registers for data and addressing. A common example of a hardware ISA is the Intel x86, which has been implemented by manufacturers in both Intel and non-Intel-based processors. Processors also vary in the ability to support or emulate different instruction sets from other processors and the ability to partition or subdivide processing capabilities to support consolidation and isolation of different workloads, operating systems, applications, and users. Other differences among processors include the number and size of internal registers and the number of bits—for example, 32-bit and the newer 64-bit processors that can work on more data in a given clock or time cycle. Processors also differ in their ability to alter speed or performance to reduce energy consumption. There are also different virtualization assist features such as Intel VT or AMD-V technology. Packaging of processors varies with the number and type of processing cores per chip, or packing, along with the amount of memory.

Processors perform calculations, run programs, and move data, all of which consume electricity. Heat is generated as a by-product of the electricity being used by a processor. Processors, particularly those with energy management and power-saving modes, vary in the amount of electricity they consume in different states of operation. Power-saving modes may involve stepping down processor clock speed or frequency, disabling functionalities that are not currently needed, or going into a or sleep or hibernation mode.

The quantity and type of processors, as well their energy consumption and subsequent heat generation, vary for different types of servers. To remove heat, some processors in larger servers have heat sinks or cooling vanes, cooling fans, or, in the case of high-end specialized processors, liquid cooling using inert coolant. Servers vary in their cooling capabilities, with most using some form of fan to remove heat and keep components cool. To experience the heat generated by a processor, place a laptop computer in your lap while doing work for an extended period of time with some background task also running that places a load on the computer. Depending on the make, model, and configuration, you may experience the heat from the laptop as well as, in a quiet environment, hear a cooling fan turning on and off as needed.

Air flow has been the most common and cost-effective means for heat removal and cooling of processors, more so with general-purpose and commercially available servers. Some servers have variable-speed cooling fans that operate at different speeds, depending on temperature, thus varying their energy consumption. Water or liquid cooling has been used in the past, particularly on older IBM and plug-compatible mainframe computers and on some supercomputers. The benefit is that water or liquid cooling is far more efficient at removing heat than air; however, the downside is cost and complexity in that electronics and water or most other liquids do not work well together. Thus, water cooling, rather than being used inside servers, has been more commonly used to remove heat from computer room air conditioning units.

With the increased density and computer power of processors that require more energy, more heat needs to be removed than ever before. Servers with the same physical footprint have, in general, increased from an electrical power density of a few kilowatts to today's densities of 6–10 kW and will soon surpass 40–50 kW per cabinet or rack. With each increase in power consumption, more heat needs to be removed, and, to reduce energy consumption, the heat needs to be removed more efficiently. The most effective location for removing heat is as close to the source as possible. While processor and chip manufacturers continue to work on more energy-efficient chips, heat still needs to be removed.

In an effort to remove heat more efficiently, liquid cooling methods involving water and other liquids are reappearing. These approaches range from specialized closed cooling systems installed directly inside server cabinets to using liquid-based cooling techniques to keep server components cool. For example, instead of immersing processors and computer cards in inert liquids, as was done in the past with some supercomputers, new techniques include placing microcoils that attach directly to chips for cooling purposes.

Using techniques such as pumped nonwater liquids that expand to gas state if exposed to room-temperature air, lower-pressure tubing and plumbing can be used for directed and pinpoint cooling. Initially this approach is being used to supply coolant to in-cabinet, next-to-cabinet in-line, top-of-cabinet, as well as ceiling-mounted cooling systems. Moving forward, these systems can be used for circulating coolant between server blades and chassis or to attach cooling to heat sinks attached directly to circuit boards and processor chips.

Research is being done into microcooling using small plumbing tubes (about the size of a human hair) to direct coolant flow between various layers of a processor chip in order to remove heat as close to the source as possible. The intent is to maximize cooling efficiency, which, in many environments, is a major source of energy consumption. The more energy that can be saved while providing more effective and efficient cooling, the more cost savings will be seen in less energy being used, less carbon impact, or, for many environments, more electrical power to be used elsewhere to enable the addition of faster servers to sustain business growth.

Given the large amount of energy used to power and cool servers, the U.S. Environmental Protection Agency (EPA) is developing Energy Star standards for servers. By the time you read this, the first tier should have deployed and work begun on a second tier to replace the first phased specification. Energy Star is a program already in place for many consumer products, including laptop computers and other household appliances, as a means of comparing the energy efficiency of various products. Many other countries either have similar programs or base their programs on Energy Star. In addition to Energy Star for servers, the EPA is also working on Energy Star for storage, data centers, telecommunications, and other entities that consume energy. Note that Energy Star is a voluntary program and is not tied to an energy emissions or carbon trading scheme. Learn more about Energy Star and related programs at www.energystar.gov.

In addition to being available in various shapes, sizes, performance capabilities, capacities, and connectivity or packing options, servers also support different operating systems. In some cases operating systems are tied to specific hardware ISA; others are more portable or have versions that have been ported to other ISAs. Microsoft Windows® is an example that runs on a variety of different x86 ISA-based servers and hypervisors. Various UNIX and Linux versions run on propriety processors and servers as well as on standard ISA hardware platforms. Some operating systems, especially older ones, are tied to specific hardware ISA and platforms.

7.2.2 Memory

Computers rely on some form of memory, ranging from internal registers to local on-board processor Level 1 (L1) and Level 2 (L2) caches, random accessible memory (RAM), nonvolatile RAM (NVRAM), or Flash, along with external disk storage. Memory, which includes external disk storage, is used for storing operating system software along with associated tools or utilities, application programs, and data. Main memory or RAM, also known as **dynamic RAM (DRAM)**, is packaged in several ways, with a common form being dual **inline memory modules (DIMMs)** for notebook or laptop, desktop PCs, and servers.

RAM main memory on a server is the fastest form of memory, second only to internal processor or chip-based registers, L1, L2, or local memory. RAM and processor-based memories are volatile and nonpersistent in that when power is removed, the contents of memory are lost. As a result, some form of persistent memory is needed to retain programs and data when power is removed. Read-only memory (ROM) and NVRAM are both persistent forms of memory in that their contents are not lost when power is removed. The amount of RAM memory that can be installed on a server varies with the specific architecture implementation and operating software being used. In addition to memory capacity and packaging format, the speed of memory is also important to be able to move data and programs quickly and avoid internal bottlenecks. Memory bandwidth performance increases with the width of the memory bus in bits and frequency in megahertz. For example, moving 8 bytes on a 64-bit bus in parallel at the same time at 100 MHz provides a theoretical 800 MByte/sec speed.

To improve availability and increase the level of persistence, some servers include battery-backed RAM or cache to protect data in the event of a power loss. On some servers, another technique to protect memory data is *memory mirroring,* in which twice the amount of memory is installed and divided into two groups. Each group of memory contains a copy of data being stored so that in the event of a memory failure beyond those correctable with standard parity and error-correction code, no data is lost. Although they are fast, RAM-based memories are also expensive and are generally used in smaller quantities compared to external persistent memories such as magnetic hard disk drives, magnetic tape, or optical-based memory media.

Figure 7.5 shows a tiered memory model that may look familiar, as the bottom portion is often expanded to show tiered storage. Looking at Figure 7.5, at the top of the memory pyramid is high-speed processor memory, followed by RAM, ROM, NVRAM, and FLASH as well as many different forms of external memory (storage—see Chapter 8).

Disk storage solutions are slower but lower in cost than RAM-based memories. They are also persistent; that is, as noted earlier, data is retained

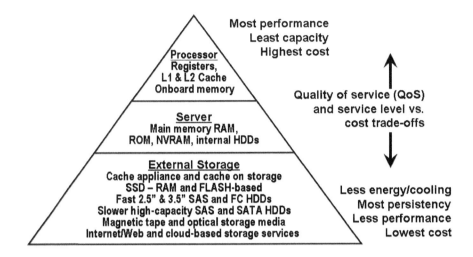

Figure 7.5   Memory and Storage Pyramid

on the device even when power is removed. As shown in Figure 7.5, less energy is used for power storage or memory at the bottom of the pyramid than at the upper levels, where performance increases. From a PCFE perspective, balancing memory and storage performance, availability, capacity, and energy to a given function, quality of service, and service-level objective for a given cost needs to be kept in perspective. It is not enough to consider only the lowest cost for the most amount of memory or storage. In addition to capacity, other storage-related metrics that should be considered include percent utilization, operating system page faults and page read/write operations, memory swap activity, and memory errors.

The difference between base 2 versus base 10 numbering systems can account for some storage capacity that appears to "missing" when actual storage is compared to what is expected. Disk drive manufacturers use base 10 (decimal) to count bytes of data, while memory chip, server, and operating system vendors typically use base 2 (binary) to count bytes of data. This has led to confusion when comparing a disk drive measured in base 10 with a chip memory measured in base 2—for example, 1,000,000,000 (10^9) bytes versus 1,073,741,824 (2^{30}) bytes. Nomenclature based on the International System of Units (SI) uses MiB, GiB, and TiB to denote millions, billions, and trillions of bytes for base 2 numbering, and MB, TB, and GB, respectively, for the same values in base 10. Most vendors document how many bytes, sometimes in both base 2 and base 10, as well as the number of

512-byte sectors supported on their storage devices and storage systems, though the information may be in the small print.

7.2.3   I/O Connectivity for Attaching Peripheral Devices

Servers support various types of connectivity for attaching peripheral devices, such as LAN (local area network) networking and I/O components for video monitors, keyboards, mice, and storage devices. Servers are continuing to evolve from proprietary host bus adapter and bus interconnects to open-standard interfaces, with the most popular currently being the **peripheral component interconnect (PCI)**. Even where servers continue to use proprietary internal I/O and communication busses and interconnects, most servers use or provide some form of PCI-based connectivity. For example, the venerable IBM mainframe leverages PCI-based FICON/Fibre Channel adapter cards via attachment to proprietary interfaces for I/O connectivity.

There are several generations of PCI technologies (see www.pcisig.com), including base PCI and its variants, PCIx, and PCIe, the current generation. PCI enables various networking and I/O devices to be attached to a computer or server—for example, PCI to 1-Gb and 10-Gb Ethernet, PCI to Fibre Channel, PCI to Fibre Channel over Ethernet-FCoE, PCI to InfiniBand, PCI to SAS or SATA, among others—using embedded on-board chips or external removable adapters and mezzanine cards in the case of blade centers. Other peripheral attachments and connectivity found on servers include IDE and ATA for older systems, USB 2.0 and 1394 Fire wire, keyboard, video, and mouse or PS2 ports, wireless Bluetooth, WiFi, WiMax, 3G, and wireless USB, among others.

The quantity and type of I/O connectivity varies with the type of server. Similar to memory and processors becoming faster while reducing their physical footprint, I/O connectivity adapters and chips are also getting smaller. I/O interconnect adapters are also being improved in terms of their capabilities to perform multiple functions. For example, networking cards and chips can also support storage services such as iSCSI, FCoE, and NAS on an Ethernet adapter. See Chapter 9 for more discussion of I/O connectivity.

7.2.4   Cabinets, Racks, and Power Supplies

Servers are available in different shapes and sizes for different functions and in a variety of packing options. Some servers are preconfigured in vendor-supplied cabinets that include cooling fans, power distribution units, or optional standby power such as battery-backed UPS systems. Cabinets can be closed with locking doors for security and to maximize cooling, as some cabinets have self-contained closed cooling systems. Servers can also be free-standing, such as tower or pedestal servers, or they may be rack-mounted in open-air equipment racks. Servers may be installed in the same cabinet as storage devices or kept separate. Other options for server cabinets include stabilizers and floor mounts for earthquake-sensitive areas or for especially tall racks. For extra-large cabinets or cabinets that contain heavy equipment, the weight of the cabinet may need to be spread over a larger area. For servers that will be installed in or near work areas, noise is another consideration, with in-the-cabinet cooling being quieter than open-based cooling systems, though this solution may be more expensive.

Servers set in a rack or cabinet may have a slide-out tray for a common keyboard and flip-up video monitor and mouse in addition to networking ports. Using networking interfaces, servers can communicate with management servers and monitoring systems for error and fault notification and response. For example, in the event of a power failure or temperature rising above a safe level, servers can be instructed to power down proactively. Keyboard-video monitor-mouse (KVM) connections can be shared across multiple servers to reduce space and cost as well as lower the number of consoles, management systems, services processors, and monitors required, and also to reduce the amount of power consumed.

Power supplies are receiving significant attention as a means of reducing energy consumption. The 80 Plus Initiative is an industry group focused on promoting power supplies that are 80% efficient or better in using energy. Power supplies use incoming AC energy, transform it to DC, and condition it for internal computer or server needs at various voltages and amperes. Power supplies often include cooling fans to keep the power supply as well as server components cool. If power supplies operate at 80% efficiency or higher, more energy is used for useful work and less energy is wasted.

Many vendors use a logo to indicate 80 Plus efficiency and document actual power supply efficiency in addition to watts or kilovolts used for nominal or normal operations as well as for start-up, peak surge, and low-power modes. Some servers, especially high-density blade servers being deployed in environments with large scaling needs, use DC power directly, without AC-to-DC conversion. DC power requires a different physical power infrastructure than the normal AC power found in most data centers; however, for some environments, DC power is worth keeping an eye on as an emerging technology.

7.2.5   Measuring and Comparing Server Performance

Comparing server performance can be like trying to compare apples to oranges. Significant differences can complicate comparing different operating systems, applications, and hardware platform configurations. Another factor is what metrics are being used—percent of CPU processor utilization, response time, number of transactions, messages, compute operations, I/O operations, files or videos served per second, or number of mega- or gigahertz. The effective performance of a server, however, is determined by how all of the components actually Work together under different loading or work conditions. When looking at servers, consider performance at low power setting modes as well as energy and cooling requirements during normal and heavy processing times.

Servers can be measured in terms of how much energy they consume in watts or kilovolts and in the amount of heat dispersed in Btu. Server power consumption or heat emission values are often provided on a name plate as average, peak, or maximum circuit requirements that may not reflect actual energy usage under different conditions. There are many workload comparison and benchmarking tools, including IOmeter, SPEC, TPC, and Microsoft ESRP, among others. However, the best test and comparison tool is whatever most closely resembles a given application and workload that the server will be supporting.

7.3   Types, Categories, and Tiers Of Servers

Laptops, desktops, and workstations, known collectively as personal computers or PCs, continue to see broad adoption from consumer to enterprise environments. These self-contained, relatively low-cost computers function stand-alone or via networks interacting with other servers. Many data centers exist in part to service and support applications and data that are accessed and relied on by distributed computing devices. Given the sheer number of desktops, workstations, and laptops in most organizations, any improvement in energy efficiency, productivity, and simplicity of management or cost will have a significant impact on the business's bottom line.

Small, floor-standing pedestals, towers, under-desk or desktop servers, as well as 1U and 2U rack servers, represent a large volume of the servers commonly deployed. Generically, these types of servers, including x86-based Intel, AMD, and IBM processors as well as other vendor-propriety ISA processors, are sometimes referred to as volume servers. They are called volume servers because their relative low cost, performance, and capabilities have led to large-volume deployments. Another type of server is a larger version that supports more memory, perhaps more processors or processing cores, more I/O and networking connectivity, along with other features such as additional cooling or standby power capability. For example, a smaller floor-standing or rack-mounted server may have from one to four processors or from 1 to 16 cores (up to four cores per processor), 2 to 16 memory DIMM slots, some number of internal disk drives, a couple of integrated I/O ports via the motherboard or main circuit board networking and KVM, I/O ports for USB, Ethernet and SAS/SATA along with some number of external PCIe or PCIx slots. Power supplies and cooling fans also vary with physical size.

Medium-sized floor-standing, pedestal, or rack-mounted servers may have more processors than smaller units, along with more memory, integrated networking, and external I/O expansion slots. Larger servers may rely on internal or external disk storage as well as networked attached and storage area networked attached storage. Large floor-standing and large-scale servers may have upwards of 128 processors each with some number of cores, large amounts of memory, some internal disks, and extensive I/O and expansion slot capabilities. Some larger servers require multiple cabinets to house expansion components and connectivity. Larger servers may be referred to generically as mainframes.

7.3.1   Blade Servers and Blade Centers

Blade servers (Figure 7.6) are a popular approach to packaging multiple individual servers in a smaller, denser enclosure while sharing some common infrastructure items including power, cooling, and KVM connectivity. They also help address PCFE issues. Uses for blade centers include migration or consolidation on a one-for-one basis from existing larger factor servers to a server blade and consolidation of underutilized physical servers to virtual machines on a server blade in a blade center. For example, applications that need their own server for performance or other purposes can be allocated their own server blade in a blade center, while other servers are consolidated onto different virtual machines running on one or more physical server blades.

Often the term *blade server* is used to describe both the blade center chassis as well as the individual server blades, which can lead to some confusion. Blade centers are the combination of some number of server blades, I/O networking and connectivity blades, optional storage modules, power supplies, and cooling fans in a chassis. Depending on the specific vendor implementation, multiple blade center chassis can fit into a single cabinet with multiple server blades in each blade chassis. A blade server, which is server blades in a blade center, can be used for a mix of consolidation of workloads as well as to reduce the physical footprint of servers while boosting performance and enabling scaling. Besides being used for production environments, blade servers can also be used in testing and in business continuity/disaster recovery environments to allow for more granular allocation of physical server resources in a reduced footprint. Another benefit of server blade centers is the modularity from being able to use various types of server blades (from the same vendor and in the same blade family) and networking and I/O connectivity options to meet specific needs.

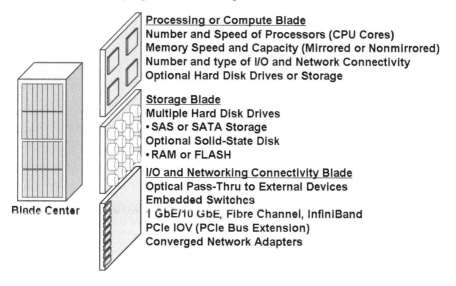

Figure 7.6   Blade Center with Various Types of Blades

Blade servers and virtual servers are often viewed as complementary solutions. For environments or applications where consolidation of smaller applications or servers is not the focus, blade centers provide a scalable and modular approach for clustering to support scale-up, scale-out, and other purposes. Scale-up and scale-out computing may mean using a blade center as a platform to run storage clustering and parallel file systems for high-performance systems, or extreme file serving for Web 2.0, cloud, NAS, and file server consolation and other applications where a large amount of storage and data needs to be online and accessible at low cost.

Blade centers and blade servers allow rapid deployment and reprovisioning for different applications and workloads in a consolidated, high-density footprint without having to physically move servers and I/O connections. Examples include virtualizing a server blade for consolidation purposes or to meet performance needs and allocating a blade to a given application that either has outgrown a virtual machine on a shared server or one that has seasonal peak workload needs. Another use is for environments whose compute and processing activities change during different phases of a project. For example, video or entertainment processing requires large amounts of computing power during certain phases. At these times, more blades can be allocated to running applications; in later phases, when more storage I/O performance and file serving are required, blades can be reconfigured to run different software to enable file and data sharing or perform other functions.

In some configurations a server blade center can be installed in the same rack or cabinet as a network blade switch and a storage subsystem. This is sometimes referred to as a "data center in a box" and may be used in smaller environments or for dedicated applications. Larger environments may devote entire racks or cabinets to multiple blade centers.

Many blade server vendors provide power calculators on their websites as well as other tools to help determine electrical power savings that a blade server can provide. Power consumption can be reduced with physical consolidation, though it may increase within a single standard 2-foot by 2-foot footprint. What this means is that implementing one or more blade centers may allow the total server footprint to be reduced, because fewer standard footprints are required. Note, however, that this may result in more weight and power in a specific footprint as a result of the denser configuration.

Server blades are available with various options, similar to traditional rackable and stackable servers, including number and type of processor cores; amount and speed of memory; types, capacities, and quantities of disk drives; and I/O connectivity options. Blade server interconnects, including embedded and via mezzanine (daughter) cards, include 1-Gb and 10-Gb Ethernet, InfiniBand, 4-Gb and 8-Gb Fibre Channel. Blade centers support pass-thru for attachment of blade servers to external switches along with support for blade switches. Blade switches vary by vendor and include Fibre Channel, 1-Gb and 10-Gb Ethernet, as well as InfiniBand from various switch manufacturers. Blade switches have some ports facing in toward the server blades and other ports facing external functioning, similar to a traditional "top of rack" or "end of row" switch up-link port.

General blade considerations include:

- KVM for managing the blades, along with CD/DVD and USB port capabilities

- Networking and storage I/O connectivity options and coexistence criteria

- Types of server processors, number of cores, and I/O capabilities per blade

- Which operating systems are supported on the different server blades

- Adequate power and cooling for a fully configured blade center

7.3.2 Virtual Servers

Virtualization is not new technology for servers. Virtualization has been around for decades, with PC emulation, logical partitions known as LPARs for dividing up a single physical server into multiple logical shared servers, and hypervisors, as well as virtual memory and virtual devices, available in proprietary form. What is new and different is the maturity and robustness of virtualization as a technology, including broad support for Intel x86-based hypervisors and other proprietary hardware ISA-based solutions.

There are many facets and functionalities of server virtualization that vary by specific implementation and product focus. Primary functionalities that various virtualization solutions can support in different combinations include:

- Emulation—coexistence with existing technologies and procedures
- Abstraction—management transparency of physical resources
- Segmentation—isolation of applications, users, or other entities
- Aggregation—consolidation of applications, operating systems, or servers
- Provisioning—rapid deployment of new servers using predefined templates

Server virtualization can be used to:

- Support and enable server consolidation to improve resource utilization
- Address PCFE issues and costs to support new applications and sustain growth
- Enable faster backup and data protection to enhance business continuity/disaster recovery capabilities
- Eliminate vendor lock-in and lower hardware costs and operating costs
- Enable transparent data movement and migration for faster technology upgrades
- Facilitate scaling beyond the limits of a single server and enable dynamic load balancing
- Improve application response time performance and user productivity
- Reduce complexity or streamline IT resource management

Server virtualization can be implemented in hardware or assisted by hardware, implemented as a stand-alone software running bare metal with no underlying software system required, as a component of an operating system, or as an application running on an existing operating system.

Figure 7.7 shows on the left a nonvirtualized model with four servers, each with an operating system and applications with access to shared or dedicated storage.

In the top middle of Figure 7.7, a consolidated server model is shown with a single server running virtualization software to create virtual

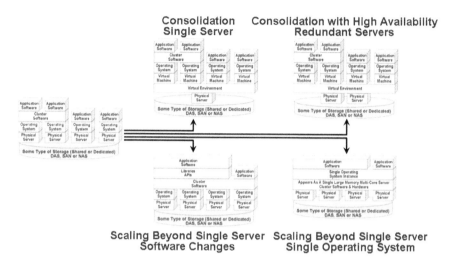

Figure 7.7   The Many Faces of Virtualization, Including Scaling with Clustering

machines. Each VM replaces a physical machine, with the VMs sharing the underlying physical hardware resources. Each VM has its own operating system and applications consolidated to a single physical server where CPU, memory, and I/O resources are shared. Also shown in this example is clustering software (such as Microsoft MSCS) installed on two VMs to provide failover capability.

The top right of Figure 7.7 shows a variation of the previous example, with two physical servers for high availability. Instead of consolidating down to a single server and introducing a single point of failure (the server), a second server is configured for both availability and redundancy as well as for performance load balancing. Since this example has two physical servers, the cluster software being used on two of the VMs could be configured with one cluster node being active on one VM and the other cluster node passive or in standby mode on the other VM.

Not all servers and applications can be consolidated. Some applications need more server resources than a single server can provide. The lower middle of Figure 7.7 shows an example of virtualization combined with other technology to enable a group of servers to host separate operating systems working together in a cluster to boost application performance. The challenge is that an application needs to be modified or designed to use specialized libraries and runtime routines or to be segmented into pieces to be able to run in parallel on different processors at the same time.

The lower right of Figure 7.7 shows a variation on the previous example, with a single operating system running on top of an abstraction technology layer that presents all underlying hardware as a single unified system. This eliminates the need for multiple operating systems to coordinate activity, including file and lock management across different nodes or requiring application modifications. This approach could be used for applications requiring extreme scaling beyond the limits of a single processor or server, such as large-scale video severing, entertainment, Web 2.0 or clustered storage file serving for cloud-based services.

Questions to consider before deploying server virtualization include:

- What are application requirements and needs (performance, availability, capacity, costs)?
- What servers can and cannot be consolidated?
- Will a solution enable simplified software management or hardware management?
- Will a solution be workable for enabling dynamic application and resource management?
- How will the technology work with existing and other new technologies?
- How will scaling performance, capacity, availability, and energy needs be addressed?
- Who will deploy, maintain, and manage the solution?
- Will vendor lock-in be shifted from hardware to a software vendor?
- How will different hardware architectures and generations of equipment coexist?
- How will the solution scale with stability?

A common use for server virtualization is consolation of underutilized servers to lower hardware and associated management costs and energy consumption along with cooling requirements Various approaches to consolidation are possible (see Figure 7.8). For example, a server's operating system and applications can be migrated as a guest to a VM existing in a virtualization infrastructure. The VMs can exist and run on a virtualization infrastructure, such as a hypervisor, that runs bare metal, or natively on a given

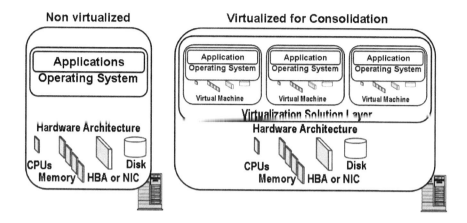

Figure 7.8   Nonvirtualized and Virtualized for Consolidation

hardware architecture, or as a guest application on top of another operating system. Depending on the implementation, different types of operating systems can exist as guests on the VMs—for example, Linux, UNIX, and Microsoft Windows all coexisting on the same server at the same time, each in its own guest VM.

VMs are a virtual entities represented by a series of data structures or objects in memory and stored on disk in the form of a file (see Figure 7.9). Server virtualization infrastructure vendors use different formats for storing the virtual machines, including information about the VM itself, configuration, guest operating system, application, and associated data. When a VM is created, for example, in the case of VMware, a VM virtual disk (VMDK) file is created that contains information about the VM, an image of the guest operating system, associated applications, and data. A VMDK can be created by converting a physical server to a VM where a source image of the physical server's operating system installation and configuration, boot files, drivers, and other information along with installed applications are created and mapped into a VMDK.

The hypervisor essentially creates multiple virtual servers, each with what appears to be a virtual CPU or processor, complete with registers, program counters, processor status words, and other items found in the hardware instruction set architecture being emulated. The theory is that with a full implementation and emulation of the underlying hardware resources being shared, guest operating systems and their applications should be able to run transparently. The reality is that, as usual, the devil may be in the

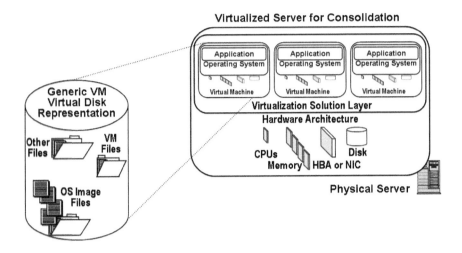

Figure 7.9   Virtual Servers and Virtual Disk File Representation

details of the specific version of the hypervisor and virtualization infrastructure, hardware firmware versions, operating system type and version, as well as application-specific dependencies on underlying operating system or server features. It is important to check with virtualization, server, operating system, and application vendors for specific supported configurations and compatibility charts as well as recommend best practices.

With a hypervisor-based VM, the VM presents to the guest operating system what appears to be a CPU, memory, I/O capabilities including LAN networking and storage, and KVM devices. Hypervisors such as VMware include virtual network interface cards (NICs), host bus adaptors (HBAs), and a virtual LAN switch all implemented in memory. The virtual LAN switch is used by the virtual NICs to enable the VMs to communicate using IP via memory instead of via a traditional physical NIC and LAN. A different form of hypervisor-based virtualization, known as *paravirtualization,* has guest operating systems and perhaps applications modified to take advantage of features in the hypervisor for improved performance. Although enhanced performance is a definite benefit of paravirtualization, not all operating systems and applications can support such features.

Popular hypervisors, including those from Microsoft, Virtual Iron, VMware, and Xen, provide emulation and abstraction of x86-based hardware instruction sets. Other vendors support other hardware instruction sets and enjoinments. For example, the IBM Z-series mainframe supports logical partitions (LPARs) and virtual machines for existing legacy mainframe operating systems and applications as well as for Linux; however, Windows and other x86-based guests and associated applications are not supported. Conversely, ports or emulations of IBM mainframe operating systems exists on x86-based systems for development, research, marketing, training, and other purposes.

For underutilized servers, the value of consolidation is sharing a server's CPU, memory, and I/O capabilities across many different VMs, each functioning as if it were a unique server, to reduce PCFE and associated hardware costs of dedicated servers. Utilization of a server hosting many VMs via a virtualization infrastructure is increased, making other servers surplus or available for redeployment for other uses. Servers that are busy during the daytime and idle during evening hours can be migrated to a VM. Then, during daytime the VM and guest operating system and application are migrated to a dedicated server or blade in a blade center to meet requirements. In the evening or during off-hours, or during seasonal periods of inactivity, the VM can be migrated to another physical server where other VMs have been consolidated, enabling some servers or blades either to be powered off or to be put into a low-power mode.

Overconsolidating, or putting to many VMs on a given server, can, as noted before, lead to resource contention, performance bottlenecks, instability that negatively impacts availability, and potentially creation of a single point of failure. A single point of failure results when, for example, eight servers are consolidated onto a single server; if that single server fails, it now affects eight VMs and all their guest operating systems, applications, and therefore users. Consequently, a balancing act is necessary among performance, availability, capacity, and energy consumption while also addressing PCFE issues.

Figure 7.10 shows another possibility: moving applications that run on the same operating system to different containers, zones, domains, or partitions of a larger server running the same operating system. In this scenario, the operating system enables applications and users to be isolated from each other while running the same version or, depending on the implementation, a different version of the same operating system.

Depending on the hardware and software configuration, underlying hardware resources such as CPUs, memory, disk or networking, and I/O adapters may be shared or dedicated to different partitions. Figure 7.10 shows a single operating system with three separate partitions (zones or containers) where different applications are running and isolated from each

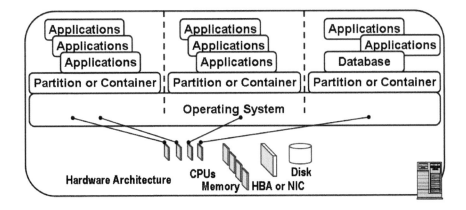

Figure 7.10   Server and Operating System Containers and Partition

other. For performance and quality of service, memory is allocated and shared across the different partitions. In this example, the server is a four-way processor with four discrete CPUs: two allocated to the leftmost partition and one each to the middle and rightmost partitions.

A hypervisor may also support different guest operating systems that run on the emulated hardware instruction set. Operating system, or hardware processor-based, partitions, containers, zones, and domains usually only support the same operating system. Operating system container-based approaches, particularly when the same version of an operating system is being used, can be vulnerable to operating system exploits and errors that can affect the entire operating system and all applications.

Another server virtualization technology is an application virtual machine that runs as a guest application on an operating system, as shown in Figure 7.11. An application virtual machine provides an abstraction or instance of a virtual machine to support a given application or environment such as, for example, Java. Applications written in Java or another language or environment capable of being run on a Java Runtime Environment (JRE) and Java virtual machine (JVM) are portable across different hardware and server environments as long as a JRE and JVM are present. For example, a JRE and JVM can exist on a laptop PC running Windows, or on a UNIX or Linux server on an IBM mainframe, or on a PDA device or cell phone. Another example of an application virtual machine and runtime environment is Adobe Flash, in which Flash-based applications are written to run on a Flash-based server.

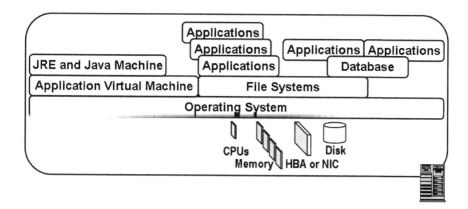

Figure 7.11   Application Virtual Machines

Although consolidation is a popular and easy-to-understand value proposition with virtualization, only a small percentage of all available servers will or can be consolidated (Figure 7.12). Not all underutilized servers are candidates for consolidation for various reasons, including application performance or quality of service requirements, and legal or regulatory requirements. Examples of servers that are not likely to be consolidated are those that support medical and emergency services or certain systems used for manufacturing regulated products. Servers may be tied to certain applications that different departments or organizations purchased and that must be kept separate for support or other reasons. It may also be necessary to keep different customers, clients, or groups of users separate for security or competitive reasons.

Server virtualization can also be used to provide management transparency and abstraction of underlying physical resources to applications and operating systems. Virtualization of servers can support massive scaling combined with clustering. For example, using server virtualization as an abstraction layer, applications that are subject to seasonal workload changes can be shifted to faster or larger servers as demand increases. Another use is to facilitate routine server maintenance including technology upgrades and replacements.

Server virtualization can be used to test new software versions or configuration changes without affecting users while reducing the number of physical servers needed. Virtualization can also aid management in achieving rapid provisioning and prestaging of new servers, operating systems, and software. For these types of applications, the value of virtualization is not consolida-

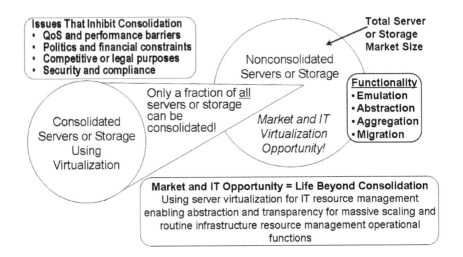

Figure 7.12   Opportunities for Server Virtualization Beyond Consolidation

tion, but to enable management transparency and abstraction of physical resources, including emulation for coexistence and investment protection.

Virtualization can also be used to support high availability, business continuity, and disaster recovery (BC/DR) by enabling proactive transparent workload migration across virtual machines while requiring fewer physical servers for recovery or standby systems. To boost performance of recovered systems that are initially supported on consolidated VMs, as new and additional server capacity is introduced into a BC/DR environment, VMs can be moved to leverage the additional servers. Another use for virtualization with regard to BC/DR is for testing of procedures and enabling training of IT staff without requiring a large number of physical servers.

Although virtualization enables consolidation of servers, server virtualization does require sufficient physical resources, including memory. Memory is required to create and support the VM as well as to host the guest operating system and applications (Figure 7.13). Unlike a physical server, whose memory is used for storing the operating system, applications, and data, a virtualized server shares memory among the different VMs, their guest operating systems, applications, and data. Depending on the virtual infrastructure implementation, additional memory may also be needed to create and maintain the VM itself, as well as a console or management system VM if applicable.

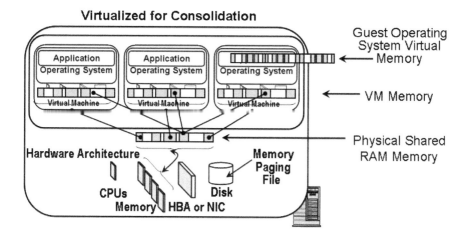

Figure 7.13   Memory Needs for Server Virtual Machines

Figure 7.13 shows at the top right an example of a guest operating system, such as Microsoft Windows, a virtual memory containing an operating system, applications, and data similar to a physical server. The virtual memory is an extension of real physical RAM memory, with some contents of RAM and other memory pages moved out to a paging disk when not needed. Also shown in Figure 7.13 on the right side is the virtual machine's memory, which is mapped to real physical RAM. Depending on the VM and virtual infrastructure implementation and operating system support, memory can be shared for common programs and data across multiple instances of the same operating system running in different VMs. For example, a read-only executable program or shared DLL library that is resident in memory may exist in one location in physical RAM yet be referenced and used by multiple VMs.

In general, more memory is better; however, the speed of the memory is also very important. Different versions and implementations of virtualization solutions support various memory configurations and limits. Check with specific vendors on their current compatibility lists for supported configurations and memory requirements. Also check with vendors for supported configurations of 32-bit and 64-bit processors; single-core, dual-core, quad-core, or eight-way processors; along with I/O cards and drivers for networking and storage devices. Note that while server consolidation can reduce hardware operating and associated costs along with power and cooling, software licensing and maintenance costs for applications and operating systems may not change unless those too are consolidated. Near-term physical hardware consolidation addresses PCFE and associated costs; however, in the longer term, additional cost savings can be obtained by addressing underutilized operating system images and application software footprints.

On virtual servers or VMs, storage is typically presented as a virtual SCSI disk. Virtual disks can be mapped to a local internal or external disk or RAID volume, and can be on an iSCSI, Fibre Channel, FCoE, or Infini-Band SAN, dedicated or shared SAS storage array as a raw LUN, or as a disk file local or on NAS. For example, VMware supports a file system called VMFS (VMware file system) for storing VMDK (VM virtual disks) with snapshots, VMotion, or live migration or other features. VMware, like other VM environments, also supports a raw device mapping (RDM) mode whereby direct access to a storage volume or file is supported, bypassing the VMFS; however, an RDM is actually a file in a VMFS that provides a transparency layer to a LUN, similar to providing a symbolic link to a LUN.

An RDM device file contains metadata, along with other information about the raw LUN or device, and optional information for support of VMotion or live migration, snapshots, and other applications. RDM can be used to minimize overhead associated with VMFS, similar to legacy applications such as databases bypassing files to perform raw and direct I/O. However, similar to traditional file systems on different operating systems, with each generation or iteration, performance improvements enable file systems performance to be on par with direct or non-file system-based I/Os while leveraging file system features including snapshots and enhanced backup.

RDM devices may be required for certain application software or IRM-related tools. These include clustering software, certain types of storage devices in clusters or to interact with storage system-based IRM utilities such as snapshot management, some types of replication, or, depending on the VM environment, to support functions such as N_Port ID Virtualization (NPIV). RDMs may not support functions such as snapshots, VCB, cloning, or other advanced functions, depending on the mode of operation.

Benefits of virtualization extend far beyond consolidation. In addition to some of the benefits we have discussed, an important capability of desktop virtualization is to facilitate various IRM-related tasks. For example, using desktop virtualization, multiple VMs can be created on a workstation, with one being active and another maintenance or recovery image. For maintenance purposes, users could be instructed to save their data, close the window for the workstation they are working on, and access a different window on the same device. While the primary VM and guest operating system along with applicable applications is being rebuilt, updated, or simply a new image copied to the VM storage locations, users can continue to work using the standby VM, which may be an older or previous version. When the upgrade is complete, users can access the upgraded software as though they had received new laptops, desktops, or workstations, without the device ever leaving them. Another benefit of using VMs on desktops and workstations is to allow certain applications to run as a guest in the background on VMs to leverage unused performance capabilities for grid and other distributed applications or services.

General server hardware and related IRM software tasks and topics include:

- Physical-to-virtual (P2V) conversion and migration
- Virtual-to-virtual (V2V) migration for maintenance or other purposes
- Virtual-to-physical (V2P) conversion and migration
- Creation of server templates for rapid provisionment
- Performance and capacity planning, problem determination, and isolation
- Volume management and file system management and configuration
- Operating system configuration and maintenance, including patches and upgrades
- Interfacing server hardware and software tools with management frameworks
- Change and configuration management
- Asset management and tracking
- Benchmarking and tuning, diagnostics, and troubleshooting
- Software upgrades, patches, and changes
- Networking, security, and storage configuration
- Backup and data protection for business continuity and disaster recovery

7.4   Clusters and Grids

Where server consolidation using virtualization can be used to reduce the number of physical servers, clustering can be used for scaling beyond the limits of a single server for performance or availability or both. Just as there are many different types of servers ranging in size, shape, capacity and performance, there are also different types of server-based clusters. Other names for server clusters include compute or server farms and grids. Clustering is often associated with high-performance scientific and research computing (see www.top500.org), but it is also very commonly deployed for availability as well as to support increased performance in everyday business and commercial applications. Clustering software is available for various platforms from server and operating system vendors as well as third-party vendors. Server clusters are available for high availability, BC/DR, load balancing, high-performance computing or I/O operations, and storage serving.

High-availability clusters can be active/passive, in which case there are two or mode nodes or servers, one active as the primary and another as a standby. If the primary server node fails or needs to be taken offline for maintenance, the standby server can assume processing responsibilities. Standby or failover clusters vary in their implementation, with some supporting application restart without loss of saved data, and other implementations enabling transparent failover and resumption without disruption to users. Some forms of clusters are active/active, with applications running actively on the different server nodes in a cluster. In the event of a failure of a server node, processing resumes on a surviving node. Depending on the type of cluster, application and transaction data may be journaled and logged so that if a failure occurs, transactions or changes can be rolled back and reapplied.

Other types of server clusters include those that are tightly coupled with propriety hardware or software and those that are loosely coupled over local and wide area networks. There are "share everything," "share some things," and "share nothing" clusters as well as local, campus, metropolitan, and wide area clusters. Clusters may serve specific functions, such as an Oracle application cluster or a parallel processing compute cluster built around specific message passing interface libraries and APIs. Some high-performance clusters allow parallel processing, with work divided up to be processed by different cluster nodes or servers at the same time. A concurrent processing cluster supports many concurrent users accessing the same application or data in situations in which several servers need to work together to provide data and application or transaction integrity.

Some vendors that offer clustered solutions use the term *grid*, apparently because "grid" seems to sound more advanced and "high tech" compared to "clusters," which have been around for decades. The differences between a cluster and a grid can lead to interesting discussions and debates, as there are various definitions and interpretations. In general, there is very little fundamental difference, because a grid can be local or remote just as a cluster can, and a grid can use off-the-shelf tightly or loosely coupled software, hardware, and networking interfaces just as a cluster can.

A grid can be an architecture, reference model, or paradigm; a service such as a utility; or a software or hardware product or solution. A way of distinguishing between a cluster and a grid might be that a grid is a collection of heterogeneous servers and operating systems, which do not work natively with each other in a trusted manner, for a common service capability. Ultimately, the differences between a cluster and grid come down to semantics, company nomenclature, and preferences. The focus should be more on what the solution is enabling and capable of and less on whether it is called a grid or a cluster.

7.5   Summary

Servers are essential for running business applications and processing data. Demand for server compute power continues to increase, and servers depend on electrical energy to function. A by-product of electrical consumption by servers that run faster and do more work in a smaller density footprint is more heat that needs to be removed. Heat removal requires additional energy. Consequently, if servers can generate less heat while doing more work, less cooling is needed, resulting in energy savings. Energy savings can be used to reduce costs or to support and sustain growth. One approach to eliminate server heat and energy consumption is to power servers off. While this is relatively easy for laptops, desktops, and workstations, simply turning all servers off is generally easier said than done. Implementation of intelligent power management, smart power switching, dynamic cooling, and other techniques, manual or automatic, can enable servers to use less energy during low-usage or idle periods.

Servers will continue to improve in processing ability while fitting into smaller footprints and improving on energy efficiency and cooling. Some of the improvements are coming from low-level fabric techniques on chips, level resources usage including power gating and variable speed or frequency, along with packing improvements.

Consolidation of underutilized servers can address PCFE issues and reduce hardware costs. However, server virtualization does not by itself address operating system and application consolation and associated cost savings. If cost savings are a key objective, in addition to reducing hardware costs, consider how software costs, including licenses and maintenance fees, can be reduced or shifted to boost savings. In the near term, there is a large market opportunity for server consolidation and an even larger market opportunity for virtualization of servers to enable scaling and transparent management on a longer-term basis.

Server hardware, software, and related clustering and tools vendors include, among others, AMD, Apple, Citrix/Xensource, Dell, Egenera, Fujitsu, HP, IBM, Intel, Lenovo, Microsoft, Novell, Racemi, Rackable, Redhat, Stratus, Sun, Supermicro, Symantec, Unisys, Verari, Virtual Iron, and VMware.

Action and takeaway points from this chapter include the following.

- Use caution with consolation to avoid introducing performance or availability problems.
- Look into how virtualization can be used to boost productivity and support scaling.
- Explore new technologies that support energy efficiency and boost productivity.
- Understand data protection and management issues pertaining to virtualization.
- Server consolation addresses hardware costs; consider software costs separately.
- Blade servers can be used for consolation as well as enabling scaling.
- Not all servers can be simply powered down, and not all redundancy can be simply removed.

- Some servers can run part of the time with lower performance and energy consumption.

- Investigate intelligent and dynamic cooling, including cooling closer to heat sources

Until the next truly revolutionary technology appears, which will be hardware that does not need software and software that does not need physical hardware, applications and virtual servers will continue to rely on physical hardware, which consumes electricity and generates heat. Watch out for having to spend a dollar to save a penny in the quest to optimize!

Chapter 8

Data Storage—Disk, Tape, Optical, and Memory

I can't remember where I stored that.

In this chapter you will learn that:

- An expanding data footprint results in increased management costs and complexity.
- Data storage management is a growing concern from a green standpoint.
- There are many aspects of storage virtualization that can help address green challenges.

Demand to store more data for longer periods of time is driving the need for more data storage capacity, which in turn drives energy consumption and consequent cooling demands. This chapter looks at various data storage technologies and techniques used to support data growth in an economical and environmentally friendly manner. These technologies also aid in sustaining business growth while building on infrastructure resource management functions, including data protection, business continuance and disaster recovery (BC/DR), storage allocation, data movement, and migration, along with server, storage, and networking virtualization topics. Although this chapter focuses on external direct attached and networked storage (either networked attached storage or a storage area network), the principles, techniques, and technologies also apply to internal dedicated storage. The importance of this chapter is to understand the need to support and store more data using various techniques and technologies to enable more cost-effective and environmentally as well as energy-friendly data growth.

8.1   Data Storage Trends, Challenges, and Issues

After facilities cooling for all IT equipment and server energy usage, external data storage has the next largest impact on power, cooling, floor space, and environmental (PCFE) considerations in most environments. In addition to being one of the large users of electrical power and floor space, with corresponding environmental impact, the amount of data being stored and the size of its the data footprint continue to expand.

Though more data can be stored in the same or smaller physical footprint than in the past, thus requiring less power and cooling, data growth rates necessary to sustain business growth, enhanced IT service delivery, and new applications are placing continued demands on available PCFE resources.

A key driver for the increase in demand for data storage is that more data is being generated and stored for longer periods of time as well as more copies of data in multiple locations. This trend toward increasing data storage will likely not slow anytime soon for organizations of all sizes.

The popularity of rich media and Internet-based applications has resulted in explosive growth of unstructured file data that requires new and more scalable storage solutions. Applications such as video pre- and post-production processing, animation rendering, on-demand video and audio, social networking websites, and digitalization of data from cell phones, personal digital assistants (PDAs) and other sources have increased burdens on storage performance and capacity. Unstructured data includes spreadsheets, PowerPoint presentations, slide decks, Adobe PDF and Microsoft Word documents, Web pages, and video and audio JPEG, MP3, and MP4 files.

The diversity of rich media and Internet applications ranges from many small files with various access patterns to more traditional large video stream access. Consequently, storage systems, in order to scale with stability to support Internet and Web 2.0 applications, will need to support variable performance characteristics from small random access of meta-data or individual files to larger streaming video sequences. Data growth rates range from the low double digits to high double or triple digits as more data is generated, more copies of data are made and, more data is stored for longer periods of time.

While structured data in the form of databases continues to grow, for most environments and applications, it is semistructured email data and

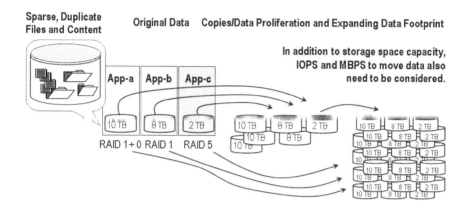

Figure 8.1   Expanding Data Footprint due to Data Proliferation and Copies Being Retained

unstructured file data that creates the biggest data footprint impact and sub-sequent bottlenecks. Unstructured data has varying input/output (I/O) characteristics that change over time, such as data that starts out with a lot of activity, then goes idle for a time before extensive reads, as in the case of a video or audio file becoming popular on a media, entertainment, social net-working, or company-sponsored website. As another example, usually, when a development or research project is completed, the data or intellec-tual property is archived or migrated to lower-cost, lower-performance bulk storage until it is needed again for further research or sequel projects.

The data footprint is the total data storage needed to support applica-tion and information needs. Your data footprint may, in fact, be larger than the actual amount of data storage you have, or you may have more aggregated data storage capacity than actual data. A general approach to determine your data footprint is simply to add up all of your online, near-line, and offline data storage (disk and tape) capacity. For example, con-sider all the data being stored at home on personal computers and laptops, PDAs, digital cameras and video recorders, TiVo sets and DVRs, USB fixed and removable disk drives, among other media that support various data and information needs.

Digital households, that is, homes with one or more computers and other electronic equipment, may have from 500 GB to over 1 TB of data. These homes' storage needs will continue to grow. The importance of understanding digital data growth needs for homes is to be able to put into scale the amount of data that needs to be stored in IT data centers to support existing and emerging applications and services.

Suppose that a business has 20 TB of data storage space that is allocated and being used for databases, email, home directories, shared documents, engineering documents, financial, and other data in different formats, both structured and unstructured. For these 20 TB of data, the storage space is probably not 100% used; database tables may be sparsely allocated, and there is likely duplicate data in email and shared document folders. However, to keep the example straightforward, assume that of the 20 TB, two complete copies are required for BC/DR purposes, and 10 TB are duplicated to three different areas on a regular basis for application testing, training, and business analysis and reporting. See Figure 8.1.

The overall data footprint is the total amount of data, including all copies plus the additional storage required to support that data, such as extra disks for redundant array of independent disks (RAID) protection or remote mirroring. In this overly simplified example, the data footprint and subsequent storage requirement amount to several times the 20 TB of data. And the larger the data footprint, the more data storage capacity and performance bandwidth are needed and that have to be powered, cooled, and housed in a rack or cabinet on a floor somewhere.

Costs associated with supporting an increased data footprint include:

- Data storage hardware and management software tools acquisition
- Associated networking or I/O connectivity hardware and services
- Recurring maintenance and software renewal fees
- Facilities fees for floor space, power, and cooling
- Physical and logical security of data and IT technology assets
- Data protection for high availability and BC/DR, including backup, replication, and archiving

It is debatable how much energy in a typical data center is actually consumed by storage (internal to servers and external) as well as how much data is active or inactive. The major power draws for common storage systems are usually spinning hard disk drives (HDDs) and their enclosures, which account for, on average, 66–75%; controllers and related I/O connectivity components generally account for most of the balance of electrical power consumption. Consequently, data storage is an important area for energy optimization and efficiency improvements.

One approach to reducing your footprint is simply to stop spending and put a cap on growth. For most environments, freezing growth is a bit draconian, but it is an option. A better approach is to do more with what you have or do more with less—that is, enable growth via consolidation and optimization to the point where further consolidation and optimization become self-defeating.

8.2   Addressing PCFE Storage Issues

There are many approaches to addressing PCFE issues associated with storage, from using faster, more energy efficient storage that performs more work with less energy, to powering down storage that is supporting inactive data, such as backup or archive data, when it is not in use. While adaptive and intelligent power management techniques are increasingly being found in servers and workstations, power management for storage has lagged behind.

General steps to doing more with your storage-related resources without negatively impacting application service availability, capacity, or performance include:

- Assess and gain insight as to what you have and how it is being used.
- Develop a strategy and plan (near-term and long-term) for deployment.
- Use energy-effective data storage solutions (both hardware and software).
- Optimize data and storage management functions.
- Shift usage habits to allocate and use storage more effectively.
- Reduce your data footprint and the subsequent impact on data protection.
- Balance performance, availability, capacity, and energy consumption.
- Change buying habits to focus on effectiveness.

- Measure, reassess, adjust, and repeat the process.

Approaches to improve storage PCFE efficiency include:

- Spin down and power off HDDs when not in use.
- Reduce power consumption by putting HDDs into a slower mode.
- Do more work and store more data with less power.
- Use FLASH and random-access memory (RAM), and solid-state disks (SSDs).
- Consolidate to larger-capacity storage devices and storage systems.
- Use RAID levels and tiered storage to maximize resource usage.
- Leverage management tools and software to balance resource usage.
- Reducing your data footprint via archiving, compression, and deduplication

Yet another approach is simply to remove or mask the problems. For example, address increased energy or cooling costs, emissions or carbon taxes if applicable, higher facilities costs if floor space is constrained, or outsource to a managed service provider, co-location, or hosting facilities. As is often the case, your specific solution may include different elements and other approaches in various combinations, depending on your business size and environment complexity.

8.3   Data Life Cycle and Access Patterns

Data has a life cycle that extends from when it is created and initially stored to when it is no longer needed and removed or deleted from storage. What varies is how long the data remains active, either being updated or in a static read-only mode, along with how the data needs to be retained when it is no longer being actively used. There are differing schools of thought on how long to keep data, from keeping as little data as possible for the shortest period of time to keeping everything indefinitely for compliance, regulatory, or internal business reasons.

Traditional life cycles may be hours or days for transaction-type data, starting with its initial creation, then the time frame when it is being actively worked with, then tapering off rapidly with only occasional read access needed. This type of life cycle lends itself to a tiered storage model and different forms of storage management, sometimes referred to as **information life-cycle management (ILM)**. ILM involves migrating unused or inactive data off online primary storage to secondary near-line or offline storage. The objectives of ILM is to reduce costs, simplifying management as well as addressing PCFE issues.

A new and emerging data life-cycle pattern for unstructured data has applications initially creating and storing data followed by a period of inactivity, lasting hours or days, leading to an active access period, followed by inactivity and the cycle continuing. Examples of this are Web 2.0-related applications such as a video on You-Tube or a popular blog, website, audio file, slide deck, or even eBooks. The data access activity picks up when the first wave of users discovers and tells others about the content, then there is a period of inactivity until a second wave discovers the information, with the cycle repeating.

Other changing data access patterns and characteristics include traditional business transactions and commercial applications with small random I/O operations per second (IOPs) and high-performance and specialized computing involving large sequential data streams. New access patterns are seeing business applications performing more I/Os, including larger sequential operations, to support rich media applications, while high-performance computing still relies on large sequential data along with growth in using more small random I/Os for meta data handling as well as processing large numbers of small files.

Given the different characteristics and application service requirements for data, different types of storage support online active data along with inactive idle data that vary in performance, availability, capacity, and energy consumption per price point and category of storage. To address the different data access and activity patterns and points in the data life cycle, virtualization provides a means to abstract the different tiers and categories of storage to simplify management and enable the most efficient type of storage to be used for the task at hand.

8.4   Tiered Storage—Balancing Application Service with PCFE Requirements

Tiered storage is an umbrella term and is often referred to by the type of HDD, by the price band, or by the architecture. Tiered storage embraces tiered media, including different types and classes of HDDs, which vary in performance, availability, capacity, and energy usage. Other storage media such as SSDs, magnetic tape, as well as optical and holographic storage devices are also used in tiered storage.

Tiered storage—various types of storage media configured for different levels of performance, availability, capacity, and energy (PACE)—is a means to align the appropriate type of IT resources to a given set of application service requirements. Price bands are a way of categorizing disk storage systems based on price to align with various markets and usage scenarios—for example, consumer, small office/home office, and low-end small to medium-size business (SMB) in a price band of under $6,000; mid- to high-end SMB in middle price bands into the low $100,000 range; and small to large enterprise systems ranging from a few hundred thousand dollars to millions of dollars.

Figure 8.2 shows examples of how tiered storage can be aligned. The lower left portion illustrates the use of high-performance HDDs and applicable RAID configurations to meet Tier 1 service needs measured on a cost-per-transaction basis. The upper right portion shows the other extreme: the

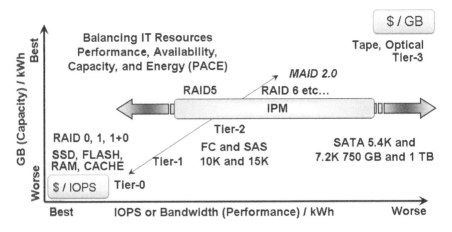

Figure 8.2   Balancing Tiered Storage and Energy to Service Needs for Active and Idle States

most capacity with lowest performance and optimum energy efficiency of offline tape and optical storage. The IT balancing act is to align a given tier of storage to specific application or data needs using PACE resources in an optimal PACE way.

Another dimension of tiered storage is tiered access, meaning the type of storage I/O interface and protocol or access method used for storing and retrieving data—for example, high-speed 8-Gb Fibre Channel (8GFC) and 10-Gb Fibre Channel over Ethernet (10FCoE) versus older and slower 4GFC or low-cost 1-Gb Ethernet (1GbE), or high-performance 10GbE-based iSCSI for shared storage access, or serial attached SCSI (SAS) and serial ATA (SATA) for direct attached storage (DAS), or shared storage between a pair of clustered servers.

Additional examples of tiered access include file- or network attached storage (NAS)-based access using network file system (NFS) or Windows-based Common Internet File System (CIFS) file sharing, among others. Different categories of storage systems combine various tiered storage media with tiered access and tiered data protection. For example, tiered data protection includes local and remote mirroring (also known as replication), in different RAID levels, point-in-time (pit) copies or snapshots, and other forms of securing and maintaining data integrity and meet data protection requirements.

8.4.1   Tiered Storage System Architectures

Tiered storage solutions, also known as different storage system architectures, include high-end cache-centric or monolithic frame-based systems typically found in upper-price-band or mid-range and clustered storage systems. Definitions of these terms have become interface-, protocol-, and host application-independent, whereas in the past there were clear lines of delineation between different storage system architectures, similar to the traditional lines of demarcation for various types of servers.

Differences used to exist between block and file (NAS)-based solutions or enterprise (mainframe) and open systems or modular and monolithic systems, but now the packaging and features/functions are blurred. Examples of high-end enterprise-class storage systems include EMC DMX, Fujitsu Eternus, HDS USP, HP XP, and IBM DS8000. All of these systems are characterized by their ability to scale in terms of performance, capacity, availability, physical size, functionality, and connectivity. For example, these systems support a mix of IBM mainframe FICON and ESCON connectivity along with open-system Fibre Channel, iSCSI, and NAS access, natively or via optional gateways, routers, or protocol converters.

In addition to supporting both open-system and mainframe servers natively, high-end cache-centric storage systems, as their name implies, have very large amounts of cache to boost performance and support advanced feature functionality. Some systems support over 1,000 HDDs including ultra-fast SSD-based devices, fast Fibre Channel HDDs, and lower-cost and high-capacity fat SATA or Fibre Channel HDDs. While smaller mid-range storage systems can in some cases rival the performance of cache-centric systems while drawing less power, an advantage of the larger storage systems can be to reduce the number of storage systems to manage for large-scale environments.

Mid-range and modular storage systems span from the upper end of price bands for enterprise solutions down to price bands for low-end SMB-based storage solutions. The characteristics of mid-range and modular storage systems are the presence of one or two storage controllers (also known as nodes), storage processors or heads, and some amount of cache that can be mirrored to the partner controller (when two controllers exist). Dual controllers can be active/passive, with one controller doing useful work and the other in standby mode in the event of a controller failure.

Mid-range and modular controllers attach to some amount of storage, usually with the ability to support a mix of high-performance fast HDDs and slow, large-capacity HDDs, to implement tiered storage in a box. The controllers rely on less cache instead of cache-centric solutions, although some scenarios that leverage fast processors and RAID algorithms can rival the performance of larger, more expensive cache-centric systems. As of this writing, modular or mid-range storage systems max out in the 250- to 400-HDD range, but by the time you read this, those numbers will have increased along with performance.

Clustered storage is no longer exclusive to the confines of high-performance sequential and parallel scientific computing or ultralarge environments. Small files and I/O (read or write), including meta-data information, are also being supported by a new generation of multipurpose, flexible, clustered storage solutions that can be tailored to support different applications workloads.

There are many different types of clustered and bulk storage systems. Clustered storage solutions may be block (iSCSI or Fibre Channel), NAS or file serving, virtual tape library (VTL), or archiving and object- or content-addressable storage. Clustered storage in general is similar to using clustered servers, providing scale beyond the limits of a single traditional system— scale for performance, scale for availability, and scale for capacity and to enable growth in a modular fashion, adding performance and intelligence capabilities along with capacity. For smaller environments, clustered storage enables modular pay-as-you-grow capabilities to address specific performance or capacity needs. For larger environments, clustered storage enables growth beyond the limits of a single storage system to meet performance, capacity, or availability needs.

Applications that lend themselves to clustered and bulk storage solutions include:

- Unstructured data files, including spreadsheets, PDFs, slide decks, and other documents
- Email systems, including Microsoft Exchange Personal (.PST) files stored on file servers
- Users' home directories and online file storage for documents and multimedia
- Web-based managed service providers for online data storage, backup, and restore
- Rich media data delivery, hosting, and social networking Internet sites
- Media and entertainment creation, including animation rendering and postprocessing
- High-performance databases such as Oracle with NFS direct I/O
- Financial services and telecommunications, transportation, logistics, and manufacturing
- Project-oriented development, simulation, and energy exploration
- Low-cost, high-performance caching for transient and look-up or reference data
- Real-time performance including fraud detection and electronic surveillance
- Life sciences, chemical research, and computer-aided design

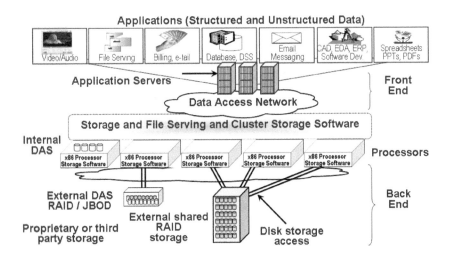

Figure 8.3   Clustered and Bulk Storage

Clustered storage solutions go beyond meeting the basic requirements of supporting large sequential parallel or concurrent file access. Clustered storage systems can also support random access of small files for highly concurrent online and other applications. Scalable and flexible clustered file servers that leverage commonly deployed servers, networking, and storage technologies are well suited for new and emerging applications, including bulk storage of online unstructured data, cloud services, and multimedia, where extreme scaling of performance (IOPS or bandwidth), low latency, storage capacity, and flexibility at a low cost are needed.

The bandwidth-intensive and parallel-access performance characteristics associated with clustered storage are generally known; what is not so commonly known is the breakthrough to support small and random IOPS associated with database, email, general-purpose file serving, home directories, and meta-data look-up (Figure 8.3).

More nodes, ports, memory, and disks do not guarantee more performance for applications. Performance depends on how those resources are deployed and how the storage management software enables those resources to avoid bottlenecks. For some clustered NAS and storage systems, more nodes are required to compensate for overhead or performance congestion when processing diverse application workloads. Other things to consider include support for industry-standard interfaces, protocols, and technologies.

Scalable and flexible clustered file server and storage systems provide the potential to leverage the inherent processing capabilities of constantly improving underlying hardware platforms. For example, software-based clustered storage systems that do not rely on proprietary hardware can be deployed on industry-standard high-density servers and blade centers and utilizes third-party internal or external storage. Clustered storage is no longer exclusive to niche applications or scientific and high-performance computing environments. Organizations of all sizes can benefit from ultrascalable, flexible, clustered NAS storage that supports application performance needs from small random I/O to meta-data lookup and large-stream sequential I/O that scales with stability to grow with business and application needs.

Additional considerations for clustered NAS storage solutions include the following.

- Can memory, processors, and I/O devices be varied to meet application needs?

- Is there support for large file systems supporting many small files as well as large files?

- What is the performance for small random IOPS and bandwidth for large sequential I/O?

- How is performance enabled across different application in the same cluster instance?

- Are I/O requests, including meta-data look-up, funneled through a single node?

- How does a solution scale as the number of nodes and storage devices is increased?

- How disruptive and time-consuming is adding new or replacing existing storage?

- Is proprietary hardware needed, or can industry-standard servers and storage be used?

- What data management features, including load balancing and data protection, exists?

- What storage interface can be used: SAS, SATA, iSCSI, or Fibre Channel?

- What types of storage devices are supported: SSD, SAS, Fibre Channel, or SATA disks?

As with most storage systems, it is not the total number of HDDs, the quantity and speed of tiered-access I/O connectivity, the types and speeds of the processors, or even the amount of cache memory that determines performance. The performance differentiator is how a manufacturer combines the various components to create a solution that delivers a given level of performance with lower power consumption. To avoid performance surprises, be leery of performance claims based solely on speed and quantity of HDDs or on speed and number of ports, processors, and memory. How the resources are deployed and how the storage management software enables those resources to avoid bottlenecks are more important. For some clustered NAS and storage systems, more nodes are required to compensate for overhead or performance congestion.

8.4.2　Tiered Storage Media or Devices

Tiered storage mediums or devices, often generically referred to as tiered storage, include different types of magnetic HDDs such as fast, high-performance Fibre Channel and SAS and lower-performing, high-capacity SATA, SAS, and Fibre Channel HDDs. Other types of tiered storage devices that can be configured into storage solutions include magnetic tape, optical media (CDs, DVDs, magneto-optical) and semiconductor disk drives (SSDs).

Given that data storage spans categories from active online and primary data to offline and infrequently accessed archive data, different types of storage media addressing different value propositions can be found in a single storage solution. For example, to address high-performance active data, the emphasis is on work per unit of energy at a given cost, physical, and capacity footprint. On the other hand, for offline or secondary data not requiring performance, the focus shifts from energy efficiency (doing more work per unit of energy) to capacity density per cost, unit of energy, and physical footprint.

Compare different tiered storage media based on what applications and types of data access they will be supporting while considering cost and physical footprint. Also consider the performance, availability, capacity, and effective energy efficiency for the usage case, such as active or idle data. As an example, a current-generation 146-GB, 15,500 (15.5K)-RPM, 4-Gb

Fibre Channel or SAS 3.5-inch HDD consumes the same, if not less, power than a 750-GB, 7,200 (7.2K)-RPM SATA or SAS 3.5-inch HDD. For active online data, the 15.5K-RPM HDD delivers more performance per unit of energy than the larger-capacity SATA HDD. However, for capacity-intensive applications that do not need high performance, the SATA drive has better density per unit of energy in the same physical footprint as the faster 146-GB HDD. Which drive to use depends on the application; increasingly, a mix of high-speed FC or SAS drives is configured in a storage system with some number of lower-performing, high-capacity or FAT HDDs for a tiered storage solution in a box.

8.4.2.1   Solid-State Devices (SSDs)

The need for more effective I/O performance is linked to the decades old, and still growing, gap between server and storage performance, where the performance of HDD storage has not kept up with the decrease in cost and increase in reliability and capacity compared to improvements in server processing power.

Reducing energy consumption is important for many IT data centers. Although there is discussion about reducing energy by doing less work or powering down storage to reduce energy use, the trend is toward doing more with less power per unit of work. This includes intelligent power management when power consumption can be reduced without compromising application performance or availability as well as doing more IOPS or bandwidth per watt of energy.

FLASH is relatively low-cost and persistent memory that does not lose its content when power is turned off. USB thumb drives are a common example. DDR/RAM is dynamic memory that is very fast but is not persistent, and data is lost when power is removed. DDR/RAM is also more expensive than FLASH. Hybrid approaches combine FLASH for persistency, high capacity, and low cost with DDR/RAM for performance.

There is a myth that SSD is only for databases and that SSD does not work with files. The reality is that in the past, given the cost of DRAM-based solutions, specific database tables or files, indices, log or journal files, or other transient performance-intensive data were put on SSDs. If the database were small enough or the budget large enough, the entire database may have been put on SSDs. Given the cost of DRAM and FLASH, however, many new applications and usage scenarios are leveraging SSD technologies.

For example, NFS filer data access can be boosted using caching I/O accelerator appliances or adapter cards.

More SSDs are not in use because of the perceived cost. The thought has been that SSDs in general costs too much compared to HDDs. When compared strictly on a cost per gigabyte or terabyte basis, HDDs are cheaper. However, if compared on the ability to process I/Os and the number of HDDs, interfaces, controllers, and enclosures necessary to achieve the same level of IOPS or bandwidth or transaction or useful work, then SSDs may be more cost-effective for a given capacity. The downside to RAM compared to HDD is that electrical power is needed to preserve data.

RAM SSDs have, over the years, addressed data persistence issues with battery-backed cache or in-the-cabinet uninterruptible power supply (UPS) devices to maintain power to memory when primary power is turned off. SSDs have also combined battery backup with internal HDDs, where the HDDs are either stand-alone, mirrored, or parity-protected and powered by a battery to enable DRAM to be flushed (destaged) to the HDDs in the event of a power failure or shutdown. While DRAM-based SSDs can exhibit significant performance advantages over HDD-based systems, SSDs still require electrical power for internal HDDs, DRAM, battery (charger), and controllers.

FLASH-based memories have risen in popularity because of their low cost per capacity point and because no power is required to preserve the data on the medium. FLASH memories have become widespread in low-end USB thumb drives and MP3 players.

The downsides to FLASH memories are that their performance, particularly on writes, is not as good as that of DRAM memories, and, historically, FLASH has a limited duty cycle in terms of how many times the memory cells can be rewritten or updated. In current enterprise-class FLASH memories, the duty cycles are much longer than in consumer or throw-away FLASH products.

The best of both worlds may be to use RAM as a cache in a shared storage system combined with caching algorithms to maximize cache effectiveness, optimize read-ahead and write-behind, and parity to boost performance. FLASH is then used for data persistence as well as lower power consumption and improved performance compared to an all-HDD storage system.

As examples, Texas Memory Systems (TMS) announced in 2007 its RAM-SAN 500, which has a RAM cache and FLASH-based storage for persistence, and EMC announced in January 2008 a FLASH-based SSD as an option in its DRAM cache-centric DMX series of storage systems. For EMC, SSD and in particular DRAM is nothing new, the company having leveraged the technologies back in the late 1980s and what ultimately became the successful DMX line of storage systems in the form of a large cache-centric storage system.

8.4.2.2   Fast or High-Performance and Fat-Capacity HDDs

As a technology, magnetic HDDs are over 50 years old, and they have evolved significantly over that time, increasing usable capacity, performance, and availability while reducing physical footprint, power consumption, and cost. The mainstays of data center storage solutions today are based on 3.5-inch high-performance enterprise and high-capacity desktop HDDs along with emerging small-form-factor 2.5-inch high-performance enterprise HDDs. With a growing focus on "green" storage and addressing power, cooling, and floor space issues, a popular trend is to consolidate data from multiple smaller HDDs onto a larger-capacity HDD to boost storage capacity versus energy usage for a given density ratio. For idle or inactive data, consolidating storage is an approach to addressing PCFE issues; however, for active data, using a high-performance drive to do more work using fewer HDDs is also a form of energy efficiency. As seen in Table 8.1, each successive generation of HDDs had improved energy usage.

Table 8.1   Balancing Performance, Availability, Capacity, and Energy Across Different HDDs

Capacity	73 GB	73 GB	73 GB	146 GB	300 GB	500 GB	750 GB	1.5 TB
Speed (RPM)	15.4K	15.4K	15.5K	15.5K	15.5K	7.2K	7.2K	7.2K
Interface	2GFC	4GFC	4GFC	4GFC	4GFC	SATA	SATA	SATA
Active watts/hour	18.7	15.8	15.2	17.44	21.04	16.34	17.7	17.7
Capacity increase	N/A	N/A	N/A	2×	4×	6.5×	10×	20×

How much energy is needed to power 100 TB of storage? The answer depends on the size of the storage system (e.g., price band) if performance or capacity is optimized, the category or type of tiered storage medium being used, how it is configured, and, if comparing raw versus usable RAID-protected storage, which RAID level plus storage is being compared on an effective basis using a compression or de-duplication ratio.

In Table 8.1, active watts/hour represents an average burdened configuration, that is, the HDD itself plus associated power supplies, cooling, and enclosure electronics per disk for active running mode. Lower power consumption can be expected during low-power or idle modes as well as for an individual disk drive minus any enclosure or packaging. In general, 100 TB of high-performance storage will require more power and therefore cooling capacity than 100 TB of low-cost, high-capacity disk-based storage. Similarly, 100 TB of high-capacity disk-based storage will consume more power than 100 TB of magnetic tape-based storage. As an example, a mid-range, mid-price storage system with redundant RAID controllers and 192 750-GByte, 7,200-RPM or 5,400-RPM SATA HDDs in a single cabinet could consume about 52,560 kWh of power per year (not including cooling). Assuming an annual energy cost of 20 cents per kilowatt-hour, which factors cost to cool the 100 TB of storage along with applicable energy surcharges or perhaps even carbon taxes being passed on by an electric utility, the total energy cost would be about $10,500. This works out to about 39.4 tons (this will vary by specific location and type of energy being used) of CO_2 emissions per year or the equivalent of about five to eight automobiles (which, of course, would vary depending on the type of automobile, miles driven per year, and type of fuel being used for the vehicle). To put this even more into perspective, for the 100 TB in this scenario, at 20 cents, the cost to power a terabyte of storage is about a penny an hour.

The previous is a middle-of-the-road example, with tape-based solutions being much lower in cost and high-end, large-enterprise storage systems with many fast HDDs being more expensive. "Your mileage will vary" is a key phrase, as different vendors with similar configurations and will have different power consumption levels, based on the overall energy efficiency of their systems independent of data footprint reduction or other enhancements capabilities.

Large-capacity drives certainly store more data at less power per gigabyte. This comes, however, at the expense of reduced performance, which can be aggravated due to density. Table 8.1 shows several different types of HDDs, ranging from 73 GB to 1.5 TB. The power consumption values shown in Table 8.1 may be higher than those normally found in HDD manufacturer specification sheets because they include infrastructure overhead. Overhead, or burden, includes the power used by the HDD as well as the HDD's share of the enclosure and associated cooling.

It is easy to compare drives on a power-per-gigabyte basis, but it is also important to consider the drive in the context of how it will be used. Look at efficiency and how power is used with respect to how the storage is being used. That is, if using storage for active data, look at how much work can be done per watt of energy such as IOPS per watt, bandwidth per watt for sequential, or video streams per watt of energy. If the data is inactive or idle, then consider the energy required to support a given capacity density while keeping in mind that unless it is for deep or time-delayed access, some amount of performance will be needed.

For those who think that the magnetic HDD is now dead, in actuality, just as disk is helping to keep magnetic tape around, SSD (both DRAM and FLASH) will help take some performance pressure off HDDs so that they can be leveraged in more efficient and economical ways, similar to what disk is to tape today. While magnetic HDDs continue to decline in price per capacity, FLASH price per gigabyte is declining at a faster rate, which makes storage using SSDs a very complementary technology pairing to balance performance, availability, capacity, and energy across different application tiers.

8.4.2.3 Magnetic Tape, Optical, and Holographic Storage

For applications and environments that need the lowest-energy-consuming storage and where response time or application performance are not required, for example, offline storage, magnetic tape remains a good option and companion to HDD-based online and near-line storage systems. Magnetic tape has been around for decades in different formats and with various characteristics. While utilizing similar techniques and basic principles, over time magnetic tape has evolved along the lines of other technologies, with increased densities, reduced physical size, better reliability, faster performance, easier handling, and lower costs.

Tape today is being challenged by lower-cost, high-capacity disk-based storage such as SATA devices to perform disk-to-disk (D2D) and disk-to-disk-to-tape (D2D2T) backups and archives. This is not to say that tape is dead, as it continues to be the most cost-effective medium for long-term and offline data retention. For example, tape cartridges are capable of storing over 1 TB of native, uncompressed data. Alternative media include magneto-optical media (MOs), CDs, DVDs, and emerging holographic storage.

8.4.3   Intelligent Power Management and MAID 2.0

Intelligent power management (IPM), also called adaptive power management (APM) and adaptive voltage scaling (AVS), applies to how electrical power consumption and, consequently, heat generation can be varied depending on usage patterns. Similar to a laptop or PC workstation with energy-saving modes, one way to save on energy consumption by larger storage systems is to power down HDDs when they are not in use. That is the basic premise of MAID (Figure 8.4), which stands for Massive (or Monolithic or Misunderstood) Array of Idle (or Inactive) Disks.

MAID-enabled devices are evolving from first-generation MAID 1.0, in which HDDs are either on or off, to a second generation (MAID 2.0) implementing IPM. MAID 2.0 leverages IPM to align storage performance and energy consumption to match the applicable level of storage service and is being implemented in traditional storage systems. With IPM and MAID 2.0, instead of a HDD being either on or off, there can be multiple power-saving modes to balance energy consumption or savings with performance needs.

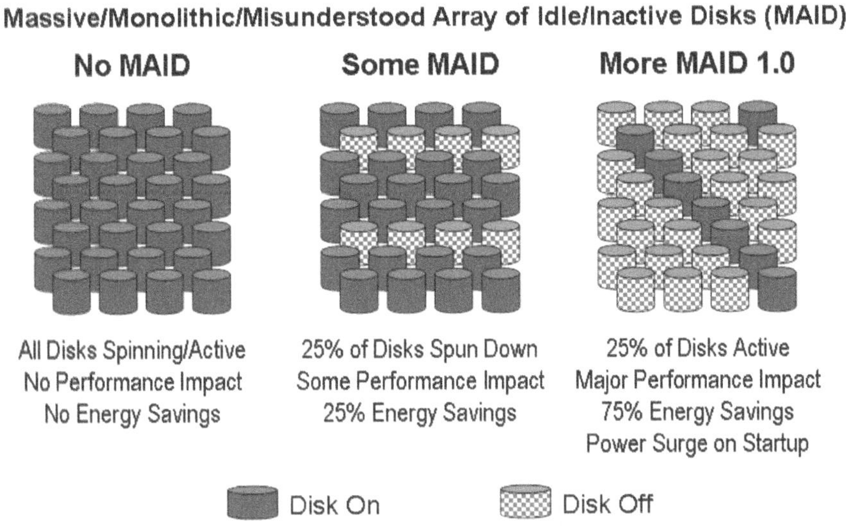

Massive/Monolithic/Misunderstood Array of Idle/Inactive Disks (MAID)

No MAID	Some MAID	More MAID 1.0

All Disks Spinning/Active
No Performance Impact
No Energy Savings

25% of Disks Spun Down
Some Performance Impact
25% Energy Savings

25% of Disks Active
Major Performance Impact
75% Energy Savings
Power Surge on Startup

Disk On   Disk Off

Figure 8.4   Intelligent Power Management and MAID

For example, a storage system can implement MAID Level 0 (no real energy savings, no impact on performance) for active data. For less active data, an administrator can choose a user-selectable setting to transition the storage system to MAID Level 1, in which power is reduced by retracting HDD read/write heads. For even more power savings, a HDD or RAID group or some other unit of storage granularity can be put into a MAID Level 2 mode, in which the speed of the drive platters is reduced. For the most power savings, the administrator can select MAID Level 3, in which a HDD or the RAID group is completely powered down or put into a suspended/standby/sleep mode. Second-generation MAID or MAID 2.0 solutions embracing IPM or other variations of adaptable power management are available from many storage vendors in general-purpose storage systems.

First-generation MAID systems were purpose-built and dedicated to the specific function of storing offline data in an energy-efficient manner by limiting the number of active disk drives at any given time. Second-generation MAID and variations of IPM implementations provide more flexibility, performance, and support for tiered storage devices while addressing PCFE concerns. For example, some vendors support RAID granularity on a RAID group basis instead of across an entire storage system. By implementing at a RAID group granularity, users are able to power down both SATA and higher-performance Fibre Channel or SAS HDDs without affecting the performance of other HDDs in other RAID groups.

Given performance or other service requirements, not all storage or data applications are appropriate for MAID. Look at second-generation MAID- and IPM-enabled storage solutions to determine the granularity and flexibility options. For example, do all of the disks in a storage system have to be under MAID management, or can some disks be included while others are excluded? Similarly, are only SATA HDDs supported, or is a mix of SATA, SAS, or even Fibre Channel disks with IPM capabilities and different MAID level granularities at the drive, RAID, or volume group level possible?

Understand the performance trade-offs. Combine MAID with other data footprint technologies including archiving, compression, and de-dupe to boost storage efficiency for online and offline data. If you are looking at secondary, online archive, or near-line storage, some of the key attributes are high-density storage (or capacity optimized), energy efficiency with intelligent power management enabled with MAID 2.0 (reduce power and cooling costs) and affordability.

8.4.4 Balancing PACE to Address PCFE Issues with Tiered Storage

Using tiered storage and aligning the most appropriate storage technology to the task or application is part of an overall approach to address power, cooling, floor-space, and environmental (PCFE) issues while balancing usage with application service-level requirements. Table 8.2 shows different tiers of storage independent of specific architectures (for example, cache-centric or monolithic frame based; modular, distributed, or clustered) or tiered access methods (DAS, SAN, NAS, or CAS), contrasting performance, availability, capacity, and energy consumption as well as relative cost.

Alignment of the most appropriate tier of storage to application needs is an effective technique, along with others such as data footprint reduction, to address data center bottlenecks without continuing the vicious cycle of sparse storage allocation and later consolidation.

From a power and performance perspective, SSD provides very good IOPS per watt or bandwidth per watt of energy used and capacity per watt. There is, however, a trade-off of cost and capacity. As an example, 1 TB of usable FLASH-based SSD can fit into 2 rack units height (2U) in a standard 19-inch rack consuming 125 W or .125 kWh of power capable of delivering 100s of MB/second bandwidth performance. The FLASH SSD is comparable to two other storage solutions: storage-centric, high-capacity, low-cost HDDs; or high-performance, medium-capacity, disk-based storage systems.

Even with the continuing drop in prices of DDR/RAM and FLASH-based SSDs and increasing capacity (Figure 8.5), for most IT data centers and applications there will continue to be a need to leverage tiered storage, including HDD based storage systems. This means, for instance, that a balance of SSDs for low-latency, high-I/O or performance hotspots along with storage-capacity, high-performance HDDs in the 146-GB and 300-GB, 15.5K-RPM class are a good fit with 500-GB, 750-GB, and 1-TB HDDs for storage capacity-centric workloads.

For active storage scenarios that do not require the ultralow latency of SSD but need high performance and large amounts of affordable capacity, energy-efficient 15K-RPM Fibre Channel and SAS HDDs provide a good balance between activity per watt, such as IOPS per watt and bandwidth per watt, and capacity, as long as the entire capacity of the drive is used to house active data. For dormant data and ultralarge storage capacity environments with a tolerance for low performance, larger-capacity 750-GB and 1.5-TB

Table 8.2   Comparison of Tiers of Storage and Service Characteristics in the Same Footprint

	Tier 0 Very High Per- formance	Tier 1 Performance and Capacity	Tier 2 Capacity and Low Cost	Tier-3 High Capacity and Low Cost
Uses	Transaction logs and jour-nal files, pag-ing files, look-up and meta-data files, very active data-base tables or indices	Active online files, data-bases, email and file serv-ers, video serving need-ing perfor-mance and storage capac-ity	Home directo-ries, file serving, Web 2.0, data backups and snapshots, bulk data storage needing large capacity at low cost	Monthly or yearly full back-ups, long-term archives or data retention with accessibility traded for cost or power savings
Comparison	Dollar per IOPSs IOPS or activ-ity per watt of energy and given data protection level	Activity per watt of energy and capacity density and given data protection level	Capacity density per energy used with perfor-mance for active data at protec-tion level	Capacity density per energy used with bandwidth when accessed at protection level
Attributes	Low capacity and high per-formance with very low power con-sumption: DDR/RAM, FLASH, or some combi-nation	Primary active data requiring availability and perfor-mance: 10K-or 15K-RPM 2.5- or 3.5-inch FC, SCSI, and SAS HDDs	Low cost, high density: 5.4K- or 7.2K-RPM SAS, SATA, or FC HDDs with capacities in excess of 1 TB	Low cost and high capacity or "FAT" 5.4K- or 7.2K-RPM SAS, SATA, or FC HDDs; mag-netic tape and optical media
Examples	Cache, SSD (FLASH, RAM).	Enterprise and mid-range arrays	Bulk and IPM-based storage	Tape libraries, MAID/IPM or optical storage, removable HDDs

"fat" HDDs that trade I/O performance for greater storage capacity provide a good capacity per watt of energy.

Another variable to consider is how the storage system is configured in terms of RAID level for performance, availability, and capacity. RAID levels affect energy consumption based on the number of HDDs or SSD (FLASH or DDR/RAM) modules being used. Ultimately, the right balance of PACE

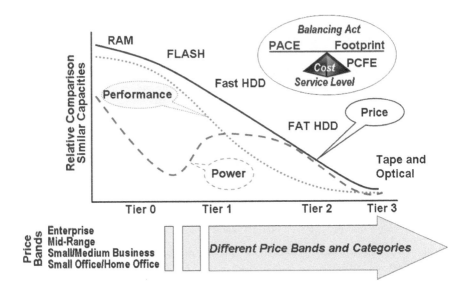

Figure 8.5   Tiered Storage Options

should be weighed with other decision and design criteria, including vendor and technology preferences, to address various PCFE issues.

8.5   Data and Storage Security

Securing stored data involves preventing unauthorized people from accessing it as well as preventing accidental or intentional destruction, infection, or corruption of information. Data encryption is a popular topic, but it is just one of many techniques and technologies that can be used to implement a tiered data security strategy. Steps to secure data involve understanding applicable threats, aligning appropriate layers of defense, and continual monitoring of activity logs, taking action as needed.

Avoid letting data security become a bottleneck to productivity, because that is a sure way to compromise a security initiative. The more transparent the security is to those who are authorized to use the data, the less likely those users will try to circumvent your efforts. A tiered data protection and security model includes multiple perimeter rings of defense to counter potential security threats. Multiple layers of defense can isolate and protect data should one of the defense perimeters be compromised from internal or external threats. Include both logical (authorization, authentication, encryption, and passwords) and physical (restricted access and locks on server, storage, and networking cabinets) security.

Data and storage security techniques and approaches include:

- Encryption of data at rest when stored on disk and when in flight over networks
- Server software, appliance, storage system, adapter, and drive-based encryption
- Key management encryption for primary and BC/DR sites
- Authentication and authorization, including rights management
- Storage logical unit number (LUN) addressing, volume zoning, mapping, and masking
- Secure management ports and management tool interfaces
- Digital shredding and secure deletion of data from retired media
- Global Positioning System (GPS) and radio frequency identification device (RFID) tracking and volume labels for removable media

8.6   Data Footprint Reduction—Techniques and Best Practices

Although data storage capacity has, in fact, become less expensive, as a data footprint expands, more storage capacity and storage management, including software tools and IT staff time, are required to care for and protect business information. By more effectively managing the data footprint across different applications and tiers of storage, it is possible to enhance application service delivery and responsiveness as well as facilitate more timely data protection to meet compliance and business objectives. To realize the full benefits of data footprint reduction, look beyond backup and offline data improvements and include online and active data.

Several methods, as shown in Table 8.3, can be used to address data footprint proliferation without compromising data protection or negatively affecting application and business service levels. These approaches include archiving of structured (database), semistructured (email), and unstructured (general files and documents) data, data compression (real-time and offline), and data de-duplication.

Table 8.3   Data Footprint Reduction Approaches and Techniques

	Archiving	Compression	De-duplication
When to use	Structured (database), semistructured (email), and unstructured	Online (database, email, file sharing), backup or archive	Backup or archiving or recurring and similar data
Characteristics	Software to identify and remove unused data from active storage devices	Reduce amount of data to be moved (transmitted) or stored on disk or tape	Eliminate duplicate files or file content observed over a period of time to reduce data footprint
Examples	Database, email, unstructured file solutions with archive storage	Host software, disk or tape, network routers, and compression appliances	Backup and archive target devices and virtual tape libraries (VTLs), specialized appliances
Caveats	Time and knowledge to know what and when to archive and delete, data and application aware	Software-based solutions require host CPU cycles, affecting application performance	Works well in background mode for backup data to avoid performance impact during data ingestion

8.6.1   Archiving for Compliance and General Data Retention

Data archiving is often perceived as a solution for compliance; however, archiving can be used for many other purposes as well, including general data footprint reduction, to boost performance and enhance routine data maintenance and data protection. Archiving can be applied to structured databases data, semistructured email data and attachments, and unstructured file data. Key to deploying an archiving solution is having insight into what data exists along with applicable rules and policies to determine what can be archived, for how long, in how many copies, and how data ultimately may be retired or deleted. Archiving requires a combination of hardware, software, and people to implement business rules.

By doing more than simply moving the data to a different location, data archiving can have one of the greatest impacts on reducing data footprint for storage in general but particularly for online and primary storage. For example, if you can identify in a timely manner what data can be removed after a project is completed, or what data can be purged from a primary database or older data migrated out of active email databases, a net improvement in application performance as well as available storage capacity should be realized.

A challenge with archiving is having the time and tools available to identify what data should be archived and what data can be securely destroyed when no longer needed. Further complicating archiving is that knowledge of the data value is also needed; this may well include legal issues as to who is responsible for making decisions on what data to keep or discard. If a business can invest in the time and software tools, as well as identify which data to archive to support an effective archive strategy, the returns can be very positive toward reducing the data footprint without limiting the amount of information available for use.

8.6.2   Data Compression (Real-Time and Offline)

Data compression is a commonly used technique for reducing the size of data being stored or transmitted to improve network performance or reduce the amount of storage capacity needed for storing data. If you have used a traditional or IP-based telephone or cell phone, watched a DVD or HDTV, listened to an MP3, transferred data over the Internet or used email, you have most likely relied on some form of compression technology that is transparent to you. Some forms of compression are time-delayed, such as using PKZIP™ to "zip" files, while others are real-time or on-the-fly, such as when using a network, cell phone, or listening to an MP3.

Two different approaches to data compression that vary in time delay or impact on application performance along with the amount of compression and loss of data are lossless (no data loss) and lossy (some data loss for higher compression ratio). In addition to these approaches, there are also different implementations of including real-time for no performance impact to applications and time delayed where there is a performance impact to applications.

In contrast to traditional "zip" or offline time-delayed compression approaches that require complete decompression of data prior to modification, online compression allows for reading from, or writing to, any location within a compressed file without full file decompression and resulting application or time delay. Real-time compression capabilities are well suited for supporting online applications including databases, online transaction processing (OLTP), email, home directories, websites, and video streaming, among others, without consuming host server CPU or memory resources or degrading storage system performance. In many cases, the introduction of appliance-based real-time compression provides a performance improvement (acceleration) to I/O and data access operations for database, shared files, Web servers as well as Microsoft Exchange Personal Storage (PST) files located in home directories.

A scenario for using real-time data compression is for time-sensitive applications that require large amounts of data, such as online databases, video and audio media servers, or Web and analytic tools. For example, databases such as Oracle support NFS3 direct I/O (DIO) and concurrent I/O (CIO) capabilities to enable random and direct addressing of data within an NFS-based file. This differs from traditional NFS operations, in which a file is sequentially read or written.

To boost storage systems performance while increasing capacity utilizations, real-time data compression that supports NFS DIO, and CIO operations expedites retrieval of data by accessing and uncompressing only the requested data. Applications do not see any degradation in performance, and CPU overhead off-loaded from host or client servers as storage systems do not have to move as much data. Data compression linking real-time for primary active storage can be applied to different types of data and storage scenarios to reduce data footprint with transparency.

Another example of using real-time compression is to combine a NAS file server configured with 146-GB or 300-GB high-performance 15.5K-RPM Fibre Channel or SAS HDDs to boost the effective storage capacity of active data without introducing the performance bottleneck associated with using larger-capacity HDDs. As seen earlier in this chapter, a 146-GB, 15.5K-RPM Fibre Channel HDD consumes the same amount of power as a 750-GB, 7.2K-RPM SATA HDD. Assuming a compression ratio of 5 to 1 could provide an effective storage capacity of 730 GB yet using high-performance HDDSs for active data. Of course, compression would vary with the type of solution being deployed and the type of data being stored.

8.6.3   De-duplication

Data de-duplication (also known as de-dupe, single-instance storage, commonalty factoring, data differencing, or normalization) is a data footprint reduction technique that eliminates the occurrence of the same data. De-duplication works by normalizing the data being backed up or stored by eliminating recurring or duplicate copies of files or data blocks, depending on the implementation.

Some data de-duplication solutions boast spectacular ratios for data reduction given specific scenarios, such as backup of repetitive and similar files, while providing little value over a broader range of applications. This contrasts with traditional data compression approaches that provide lower yet more predictable and consistent data reduction ratios over more types of data and application, including online and primary storage scenarios. For example, in environments where there is little to no common or repetitive data files, data de-duplication will have little to no impact, whereas data compression generally will yield some data footprint reduction across almost all types of data.

As an example, in the course of writing this book, each chapter went through many versions, some with a high degree of commonality resulting in duplicate data being stored. De-duplication enables multiple versions of the files to be stored, yet by saving only the changed or different data from a baseline, the amount of space required is reduced. Instead of having eight versions of a file, all about 100 KB in size, or 800 KB without compression, assuming that 10% of the data changes in each version of the file, the amount stored could be in the range of 170–200 KB, depending on what data actually changed and the de-dupe solution. The result is that, over time, instead of the same data being copied and backed up repeatedly, a smaller data footprint can be achieved for recurring data.

Some data de-duplication solution providers have either already added, or have announced plans to add, compression techniques to complement and increase the data footprint effectiveness of their solutions across a broader range of applications and storage scenarios, attesting to the value and importance of data compression to reduce data footprints.

When looking at de-duplication solutions, determine if the solution is architected to scale in terms of performance, capacity, and availability over a large amount of data, along with how restoration of data will be affected by scaling for growth. Other things to consider include how data is re-duplicated, such as real-time using inline or some form of time-delayed postprocessing, and the ability to select the mode of operation.

For example, a de-dupe solution may be able to process data at a specific ingest rate inline until a certain threshold is hit, and then processing reverts to postprocessing so as not to degrade application performance. The downside of post processing is that more storage is needed as a buffer. It can, however, also enable solutions to scale without becoming a bottleneck during data ingestion.

8.6.4 Hybrid Data Footprint Reduction—Compression and De-duplication

Compression and de-duplication are two popular capacity optimization and data footprint reduction techniques that can be readily deployed with quick results. De-duplication or compression implemented as a single solution yield significant savings on storage capacity, but using both technologies on the same data files should provide even greater space savings. The result should be no performance delays for online time- or performance-sensitive applications and no waiting for data to be reinflated during data recovery or restore operations.

Several techniques can be used either individually to address specific issues or in combination to implement a more cohesive and effective data footprint reduction strategy. Develop an overall data foot reduction strategy that leverages several techniques and technologies to address online primary, secondary, and offline data. Assess and discover what data exists and how it is used in order to manage storage needs effectively.

Determine policies and rules for retention and deletion of data that combine archiving, compression (online and offline), and de-dupe in a comprehensive strategy. The benefit of a broader, more holistic, data footprint reduction strategy is the ability to address the overall environment, including all applications that generate and use data as well as infrastructure resource management (IRM) or overhead functions that compound and affect the data footprint.

8.7 Countering Underutilized Storage Capacity

Debates exist about the actual or average storage space capacity utilization for open systems, with estimates ranging from 15% to 30% up to 65–80%. Not surprisingly, the lowest utilization estimates tend to come from vendors interested in promoting SRM tools, thin provisioning, or virtualization aggregation solutions. Research and experience indicate that low storage utilization is often the result of several factors, including limiting storage capacity usage to ensure performance; isolate particular applications, data, customers, or users; for the ease of managing a single discrete storage system; or for financial and budgeting purposes. In some cases, the expense to consolidate, virtualize, and manage may not be offset by the potential gains.

When consolidating storage, consider where and how storage is being allocated and active (or idle) in order to know what can be moved when and where. You can leverage newer, faster, and more energy efficient storage technology as well as upgrade storage systems with faster processors, I/O busses, increased memory, faster HDDs, and more efficient power supplies and cooling fans.

Looking at storage utilization only from the viewpoint of space capacity consumption, particularly for active and online data, can result in performance bottlenecks and instability in service delivery. A balanced approach to utilization should include performance, availability, capacity, and energy needs for a type of application usage and access requirements. When storage management vendors talk about how much storage budget they can save you, ask them about their performance and activity monitoring and reporting capabilities; you may be told that they are not needed or requested by their customers, or that these issues will be addressed in a future release.

Over time, underutilized storage capacity can be consumed by application growth and data that needs to be retained for longer periods of time. However, unless the capacity that is to be consumed by growth is for dormant data (idle data), any increase in I/O activity will further compound the I/O performance problem by sparsely allocating storage in the first place. A similar effect has been seen with servers, because they are perceived as being inexpensive to acquire and, until recently, not as much of an issue to operate, thus leading to the belief that since hardware is inexpensive, using more of it is a workable solution to an application problem.

Although there are several approaches to increasing storage capacity utilization, it is important to use a balanced approach: Move idle or infrequently accessed data to larger-capacity, lower-performing, cost-effective SATA HDDs, and move active data to higher-performing, energy-efficient enterprise Fibre Channel and SAS HDDs. Using a combination of faster HDDs for active data, larger-capacity storage for infrequently accessed data, and faster storage system controllers can enable more work to be done using fewer HDDs while maintaining or enhancing performance and fitting into existing power and cooling footprints or possibly even reducing energy requirements.

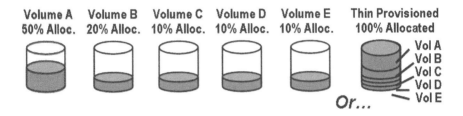

Figure 8.6   Thin Provisioning

8.7.1   Thin Provision, Space-Saving Clones

Thin provisioning is a storage allocation and management technique that presents an abstracted or virtualized view to servers and applications of how much storage has been allocated yet is actually physically available. In essence, thin provisioning, as seen in Figure 8.6, allows the space from multiple servers that have storage allocated but not actually used to be shared and used more effectively to minimize disruptions associated with expanding and adding new storage.

In Figure 8.6, each server thinks that it has, say, 10 TB allocated, yet many of the severs are only using 10% or about 1 TB of storage. Instead of having to have 5 × 10 or 50 TBs underutilized, a smaller amount of physical storage can be deployed yet thinly provisioned, with more physical storage allocated as needed. The result is that less unused storage needs to be installed and consuming power, cooling, and floor space until it is actually needed. The downside is that thin provisioning works best in stable or predictable environments, where growth and activity patterns are well understood or good management insight tools on usage patterns are available.

Thin provisioning can be described as similar to airlines overbooking a flight based on history and traffic patterns. However, like airlines overbooking a flight, thin provisioning can result in a sudden demand for more real physical storage than is available. Thin provisioning can be part of an overall storage management solution, but it needs to be combined with management tools that provide history and insight on usage patterns.

8.7.2   How RAID Affect PCFE and PACE

Redundant Arrays of Independent Disks (RAID) is an approach to addressing data and storage availability and performance. RAID as a technique and technology is about 20 years old, with many different types of hardware and software implementations available. There are several different RAID levels to align with various needs for performance, availability, capacity, and energy consumption, and cost.

Different RAID levels (Table 8.4) will affect storage energy effectiveness differently, but a balance between performance, availability, capacity, and energy (PACE) is needed to meet application service needs. In Table 8.4, for example, RAID 1 mirroring or RAID 10 mirroring and disk striping (spreading data over multiple disks to boost performance) requires more HDDs and therefore power but yield better performance than RAID 5. RAID 5 yields good read performance and uses fewer HDDs, reducing the energy footprint but at the expense of write or updating performance.

Table 8.4   RAID Levels and Their PCFE and PACE Impacts

RAID Level	Characteristics	Applications	Performance Capabilities	Energy Footprint
0	Spreads data across two or more disks to boost performance with no enhanced availability	Applications that can tolerate loss of access to data that can be easily reproduced	Very good	Very good; fewest disks
1	Data mirroring provides protection and good I/O performance with n + n disks, where n is the number of data disks	I/O-intensive OLTP and other data with high availability, including email, databases, or other I/O-intensive applications	Very good	Not good; twice as many disks needed
0 + 1	Stripe plus mirroring of data for performance and availability, n + n disks.	I/O-intensive applications requiring performance and availability	Very good	Not good; twice the disks

Table 8.4   RAID Levels and Their PCFE and PACE Impacts (continued)

RAID Level	Characteristics	Applications	Performance Capabilities	Energy Footprint
1 + 0 or 10	Similar to RAID 0 + 1, but mirrors and stripes data	I/O-intensive applications requiring performance and availability	Very good	Not good; twice the disks
3	Stripes with single dedicated parity disk, N + 1	Good performance for large, sequential, single-stream applications	Good	Good
4	Similar to RAID 3, with block-level parity protection	Using read/write cache, is well suited for file-serving environments.	Good	Good
5	Striping with rotating parity protection using n + 1 disks; parity spread across all disks for performance	Good for reads, write performance affected if no write cache; use for read-intensive data, general file and Web serving	Good for read, potential write penalty	Good; better than RAID 1
6	Disk striping with dual parity using n + 2 HDDs; reduces data exposure during a rebuild with larger-capacity HDDs	Large data capacity intensive applications that need better availability than RAID 5 provides	Good for read, potential write penalty	Very good

An effective energy strategy for primary external storage includes selecting the appropriate RAID level and drive type combined with a robust storage controller to deliver the highest available IOPs per watt of energy to meet specific application service and performance needs.

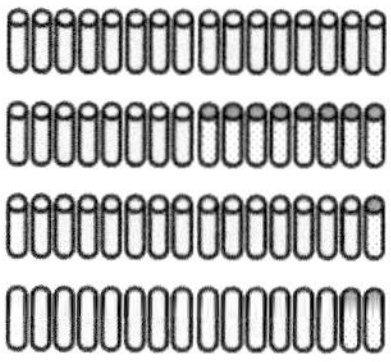

	Performance	Availability	Performance Overhead	Availability Overhead
RAID 0	Very Good	None	None	N + 0 = 0%
RAID 1	Good	Very Good	Minimum	50%
RAID 5	Poor Writes	Good	High on Write	(1P / N) 6%
RAID 6	Poor Writes	Better	High on Write	(2P / N) 12.5%

Figure 8.7   RAID Levels to Balance PACE for Application Needs

In addition to the RAID level, the number of HDDs supported in a RAID group set can affect performance and energy efficiency. In Figure 8.7, for example, N is the number of disks in a RAID group or RAID set; more disks in a RAID 1 or RAID 10 group provides more performance but with a larger PCFE footprint. On the other hand, more HDDs in a RAID 5 group spreads parity overhead across more HDDs, improving energy efficiency and reducing the physical number of HDDs. However, this solution needs to be balanced with the potential exposure of a second HDD failure during a prolonged rebuild operation. A good compromise for non-performance-sensitive applications that are storage capacity- or space-intensive might be RAID 6, particularly with solutions that accelerate parity calculations and rebuild operations.

General notes and comments about using RAID and PACE to address PCFE issues for different application and data service objectives include the following.

- Larger RAID sets can enable more performance and lower availability overhead.

- Some solutions force RAID sets to a particular shelf or drive enclosure rack.

- Balance RAID level performance and availability to type of data—active or inactive.

- Boost performance with faster drivers; boost capacity with larger capacity drives.

- With large-capacity SAS and SATA, drive capacity will affect drive rebuild times.

- Balance exposure risk during drive rebuild with appropriate RAID levels.

- Design for fault containment or isolation, balancing best practices and technology

8.8   Storage Virtualization—Aggregate, Emulate, Migrate

There are many different forms of storage virtualization, including aggregation or pooling, emulation, and abstraction of different tiers of physical storage providing transparency of physical resources. Storage virtualization can be found in server software bases, network or fabric using appliances, routers, or blades with software in switches or switching directors. Storage virtualization functionality can also be found running as software on application servers or operating systems, in network-based appliances, switchers, or routers, as well as in storage systems.

8.8.1   Volume Mangers and Global Name Spaces

A common form of storage virtualization is volume managers that abstract physical storage from applications and file systems. In addition to providing abstraction of different types, categories, and vendor storage technologies, volume managers can also be used to support aggregation, performance optimization, and enable IRM functions. For example, volume managers can aggregate multiple types of storage into a single large logical volume group that is subdivided into smaller logical volumes for file systems.

In addition to aggregating the physical storage, volume managers can perform RAID mirroring or disk striping for availability and performance. Volume managers also provide a layer of abstraction to allow different types of physical storage to be added and removed for maintenance and upgrades without affecting applications or file systems. IRM functions that are supported by volume managers include storage allocation and provisioning and data protection operations, including snapshots and replication, all of which vary with the specific vendor implementation. File systems, including clustered and distributed systems, can be built on top of or in conjunction with volume managers to support scaling of performance, availability, and capacity.

Global name spaces provide another form of virtualization by presenting an aggregated and abstracted view of various file systems. A global name space can span multiple different file systems, providing an easy-to-use interface or access view for managing unstructured file data. Microsoft Domain Name System (DNS) for Windows CIFS or Network Information Services (NIS) for NFS support global name spaces.

8.8.2   Virtualization and Storage Services

Various storage virtualization services are implemented in different locations to support various tasks. Figure 8.8 shows examples of pooling or aggregation for both block- and file-based storage, virtual tape libraries for coexistence and interoperability with existing IT hardware and software resources, global or virtual file systems, transparent data migration of data for technology upgrades, and maintenance, as well as supporting high availability and BC/ DR.

One of the most frequently discussed forms of storage virtualization is aggregation and pooling solutions. Aggregation and pooling for consolidation of LUNs, file systems, and volume pooling, and associated management, can increase capacity utilization and investment protection, including supporting heterogeneous data management across different tiers, categories, and price bands of storage from various vendors. Given the focus on consolidation of storage and other IT resources along with continued technology maturity, more aggregation and pooling solutions can be expected to be deployed as storage virtualization matures.

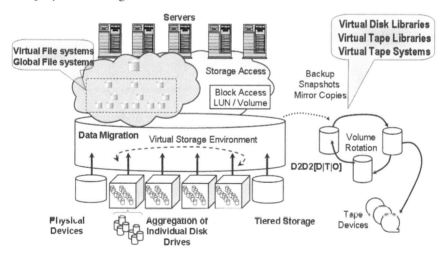

Figure 8.8   The Many Faces of Storage Virtualization

While aggregation and pooling are growing in popularity in terms of deployment, most current storage virtualization solutions are forms of abstraction. Abstraction and technology transparency include device emulation, interoperability, coexistence, backward compatibility, transition to new technology with transparent data movement and migration, as well as support for high availability and BC/DR. Some other forms of virtualization in the form of abstraction and transparency include heterogeneous data replication or mirroring (local and remote), snapshots, backup, data archiving, security, and compliance.

Virtual tape libraries (VTLs) provide abstraction of underlying physical disk drives while emulating tape drives, tape-handling robotics, and tape cartridges. The benefit is that VTLs provide compatibility with existing backup, archive, or data protection software and procedures to improve performance using disk-based technologies. VTLs are available in standalone as well as clustered configuration for availability and failover, as well as scaling for performance and capacity. Interfaces include block-based for tape emulation and NAS for file system-based backups. VTLs also support functions such as compression, de-duplication, encryption, replication, and tiered storage.

Some questions to ask regarding storage virtualization include:

- What are the various application requirements and needs?
- Will it be used for consolidation or facilitating IT resource management?
- What other technologies are currently in place or planned for the future?
- What are the scaling (performance, capacity, availability) needs?
- Will the point of vendor lock-in be shifting or costs increasing?
- What are some alternative and applicable approaches?
- How will a solution scale with stability?

Building a business case for virtual tape libraries or disk libraries to support technology transition and coexistence with existing software and procedures can be fairly straightforward. Similarly, cases for enabling transparent data migration to facilitate technology upgrades, replacements or reconfiguration along with ongoing maintenance and support can be mapped to sustaining business growth.

For example, the amount of time it takes to migrate data off older storage systems and onto newer technology can be reduced while maintaining data availability and application access, the storage resource technology can be used for a longer time, thus decreasing the time the technology is not fully utilized because of conversion and migration. Another benefit is that newer, more energy-efficient technology can be migrated in and older, less energy-efficient technology can be migrated out more quickly.

This is not to say that there are not business cases for pooling or aggregating storage; rather, there are other areas where storage virtualization techniques and solutions can be applied. This is not that different from server virtualization shifting from a consolidation role, which is the current market and industry phase, to one of enablement to support migration, maintenance, and scaling. This is true particularly for applications and workloads that are not conducive to consolidation, such as those in which more performance, capacity, or availability is needed or those that need to isolate data or customers.

Another form of storage virtualization is virtual storage servers (see Figure 8.9), or storage partitions that enable a single consolidated storage system to appear to applications and servers as multiple individual storage systems or file servers. The primary focus of virtual storage servers or partitions is to be able to isolate, emulate, and abstract the LUNs, volumes, or file systems on a shared storage server. An example is enabling a common or consolidated storage server to be shared by different applications while preventing data from being seen or accessed across applications, customers, or users.

Another possibility is to enable a LUN or volume group to provide unique addressing as well as volume mapping and masking so that different servers see what they think are unique LUNs or volumes. For example, a group of Windows servers may all want to see a LUN 0 while each needs its own unique volume so partitioning. In this case, LUN volume mapping and masking can be combined so that each server sees what it thinks is a LUN 0, yet the actual LUNs or volumes are unique. Many vendors provide some form of LUN and volume mapping and masking along with host and server partition groups or storage domains.

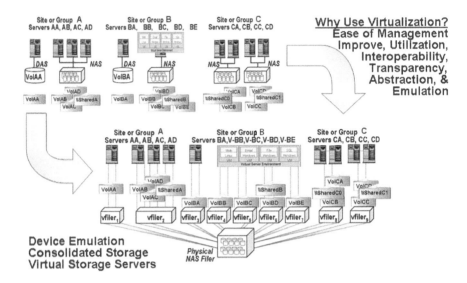

Figure 8.9 Storage Virtualization and Virtual Storage Servers or Partitions

You can also enable data to be moved or migrated to lower-cost tiers of storage for online active and reference data. Data migration can also be used to enable archiving to move data from online storage to near-line or offline storage. Lower-cost storage can be used as a target for replicating data for high availability and BC/DR, or as a target for regular data backups. Data replication and movement can be accomplished using host-based software such as volume managers or migration tools, network- or fabric-based data movers and appliances, or via storage systems.

8.9 Comparing Storage Energy Efficiency and Effectiveness

Not all storage systems consume the same amount of power or deliver the same performance with the same number of ports, controllers, cache, and disk drives. A challenge with storage systems is gauging power consumption and performance. For example, can system A use less power than system B while system A delivers the same level of performance (is it as fast), amount of capacity, or space to store data (how useful)? It is possible to create a storage system that uses less power, but if the performance is also lower than with a competing product, multiple copies of the lower-power-consuming system may be needed to deliver a given level of performance.

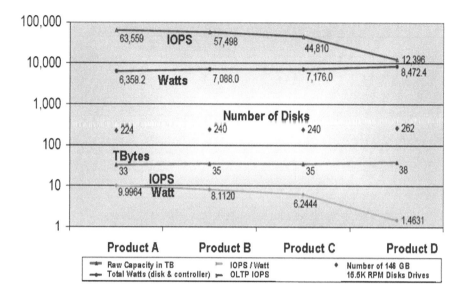

Figure 8.10   The Importance of Using Multiple Metrics for Comparison

Four different storage systems are shown in Figure 8.10, each with a similar number of the same type of high-performance HDDs performing a transaction-oriented workload. It is important to look at all the available metrics, particularly for active data storage, to see how much storage capacity will be achieved with a given density, how much power will be needed, and what level of performance will be provided, considering the total power footprint as well as the activity per energy used or IOPS per watt.

Look beyond the number of disk drives, the size or amount of cache, the number and speed of controllers to see effective capabilities. For example, different systems can have the same number of disks, but one may use less power and provide less performance, another may use the same power and give better performance, or one may use less power and offer higher performance.

Figure 8.11 shows various types of storage systems across different categories, tiers, and price bands configured to support a common baseline of 1.3 PB of raw storage. Values will from this baseline configuration depending on RAID levels, compression, de-duplication, and other optimizations. In some cases, multiple storage systems will be required to achieve this 1.3 PB of storage capacity; several scenarios are shown to help gauge footprint as well as power needs (including variable power if the systems support

Assumed kWh power cost (cents) = `0.08` <-- Enter energy cost here in cents per kWh

Configuration for 1 3PByte raw storage capacity	Non MAID Capable		MAID 1.0 or 2.0 Capable Storage			Tape Based Storage		CAS Object Storage	
	EMC CLARiiON CX3-80	NetApp R200	HDS AMS1000	Nexsan SATAbeast AutoMAID	Copan Revolution 300 MAID	Sun SL8500 LTO4 Tape	HP ESL 720e LTO4 Tape	EMC Centera 4LP	Nexsan Assureon
Number of storage systems required	4	8	4	32	2	1	2	16	32
Floor space footprint (cabinets or racks)	12	8	8	4	2	1	2	16	4
Total number disk or tape drives	1,800	2,688	1,680	1,344	1,792	32	32	1,792	1,344
Total raw non formatted capacity (Tbytes)	1,350	1,344	1,296	1,344	1,344	1,158	1,072	1,344	1,344
Total hourly bandwidth performance (Tbyte/hr)	16.14	7.20	21.89	80.64	10.41	13.82	13.82		
MAID Level 0 = Normal active disks (no saving) kWh									
MAID Level 0 - Annual Power (no cooling) kWh	474,792	598,764	606,893	210,240	128,299	18,273	20,709	560,640	236,520
MAID Level 0 - Annual energy costs ($1,000s)	$38	$48	$49	$17	$10	$1	$2	$45	$19
MAID Level 1 = Park disk read/write heads									
MAID Level 1 - Annual Power (no cooling) kWh	-	-	-	165,389	-	-	-	-	191,669
MAID Level 1 - Annual energy costs ($1,000s)	-	-	-	$13	-	-	-	-	$15
MAID Level 2 = Reduce disk RPM speed									
MAID Level 2 - Annual Power (no cooling) kWh	-	-	-	128,667	-	-	-	-	154,947
MAID Level 2 - Annual energy costs ($1,000s)	-	-	-	$10	-	-	-	-	$12
MAID Level 3 = Standby (sleep) power									
MAID Level 3 - Annual Power (no cooling) kWh	-	-	485,514	91,384	62,652	12,264	9,486	-	117,664
MAID Level 3 - Annual energy costs ($1,000s)	-	-	$39	$7	$5	$1	$1	-	$9
MAID Level-0 - kWh per Tbyte capacity (raw space)	0.040	0.051	0.053	0.018	0.011	0.002	0.002	0.048	0.020
MAID Level-1 - kWh per Tbyte capacity (raw space)				0.014					0.016
MAID Level-2 - kWh per Tbyte capacity (raw space)				0.011					0.013
MAID Level-3 - kWh per Tbyte capacity (raw space)			0.043	0.008	0.005	0.001	0.001		0.010
MAID Level-0 kWh used per Tbyte/hr bandwidth	3.358	9.493	3.165	0.298	1.406	0.151	0.171		

Figure 8.11   Performance and Energy Usage for Various Storage Solutions

either first-generation MAID or second-generation MAID 2.0 with intelligent power management or adaptive power management).

Also shown in Figure 8.11 is the performance in terabytes per hour, to show how different solutions vary in their ability to ingest and store data at a baseline configuration. From this baseline it is possible to apply various compressions and de-duplication strategies to make additional comparisons, as well as various RAID levels and RAID group sizes and numbers of hot spare disk drives.

8.10   Benchmarking

Benchmarks, particularly industry-accepted workloads that are applicable to specific applications and environments, can be a useful gauge of how different solutions compare; however, they are not a substitute for actually testing and simulating production application and workload conditions. Take benchmarks with a grain of salt when considering their applicability to your environment, application characteristics, and workloads.

8.11   Summary

IT organizations are realizing that in addition to conserving power and avoiding unneeded power usage, addressing time-sensitive applications with performance enhancements can lead to energy efficiency. Various techniques and existing technologies can be leveraged to either reduce or support growth with current power and cooling capabilities, as well as with supplemental capabilities. There are also several new and emerging technologies to be aware of and consider. These range from more energy-efficient power supplies, storage systems, and disk drive components to performance enhancements to get more work done and store more data per unit of energy consumed in a given footprint. Improvements are also being made in measuring and reporting tools to provide timely feedback on energy usage and enable tuning of cooling resources.

Deploy a comprehensive data footprint reduction strategy combining various techniques and technologies to address point solution needs as well as the overall environment, including online, near-line for backup, and offline for archive data. Using more energy-efficient solutions that are capable of doing more work per unit of energy consumed is similar to improving the energy efficiency of an automobile. Leveraging virtualization techniques and technologies provides management transparency and abstraction across different tiers, categories, and types of storage to meet various application service requirements for active and inactive or idle data. Keep performance, availability, capacity, and energy (PACE) in balance to meet application service requirements and avoid introducing performance bottlenecks in your quest to reduce or maximize your existing IT resources including power and cooling.

Action and takeaway points from this chapter include the following.

- Develop a data footprint reduction strategy for online and offline data.
- Energy avoidance can be accomplished by powering down storage.
- Energy efficiency can be accomplished by using tiered storage to meet different needs.
- Measure and compare storage based on idle and active workload conditions.

- Storage efficiency metrics include IOPS or bandwidth per watt for active data.

- Storage capacity per watt per footprint and cost is a measure for inactive data.

- Align the applicable form of virtualization for the task at hand.

Storage vendors include BlueArc, Datadomain, Dell, EMC, FalconStor, Fujifilm, Fujitsu, Gear6, Hitachi, HP, IBM, IBRIX, Imation, LeftHand, LSI, NEC, NetApp, Nexsan, Quantum, Seagate, SGI, Sun, and Xyratex, among others.

Chapter 9

Networking with Your Servers and Storage

I/O, I/O, it's off to virtual work we go.

In this chapter you will learn:

- The importance of I/O and networking with associated demand drivers and challenges
- How tiered access can be used to address various challenges
- I/O virtualization and converged data and storage networking

Like the important role that transmission networks play in access to and efficient delivery of reliable electrical power, input/output (I/O) and data networks enable IT services to be delivered to users leveraging local as well as remote servers and storage. Previous chapters have discussed the role of servers to support the processing needs of information services delivery as well as the associated power, cooling, floor space, and environmental health and safety (PCFE) impacts and issues. The amount of data being generated, copied, and retained for longer periods of time is elevating the importance of the role of data storage and infrastructure resource management (IRM). Networking and I/O connectivity technologies tie facilities, servers, storage tools for measurement and management, and best practices on a local and wide area basis to enable an environmentally and economically friendly data center. Networks play a crucial role in delivering IT services while enabling a green and virtual data center on a local as well as wide area basis. The importance of this chapter is to understand the characteristics of various physical and virtualized I/O technologies in order to align those capabilities to different usage scenarios for cost-effective IT service delivery.

9.1   I/O and Networking Demands And Challenges

To say that I/O and networking demands and requirements are increasing is an understatement. The good news is that I/O and networking in general are becoming faster, more reliable, and able to support more data movement over longer distances in a shorter timeframe at a lower cost. However, as with server and storage technology improvements, the increases in networking and I/O capabilities are being challenged by continued demand to move or access more data over longer distances in even less time and at still lower cost (see Figure 9.1).

Computers of all types rely on I/O for interaction with users directly via keyboards, video monitors, and pointing devices, along with network access to other computers and Web-based services. Another use of I/O and networking capabilities is storing and retrieval of continually growing amounts of data both locally and remotely. I/O is also important for computers to interface with storage devices for saving and retrieving stored information.

I/O and networking support user-to-computer, computer-to-computer or server-to-server, server-to-,storage as well as storage-to-storage communication on a local or remote basis. Users or clients can access IT-based services via traditional dumb terminals, small stripped-down low-cost PCs known as thin clients, personal digital assistants (PDAs), cell phones, laptops, desktop workstations, or other Internet enabled devices.

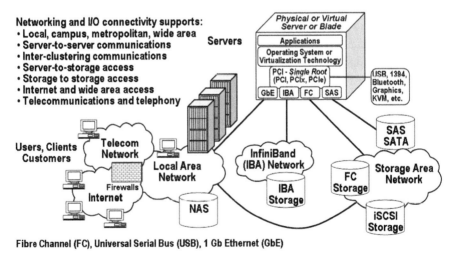

Figure 9.1   Data Center I/O and Networking—The Big Picture

In general, I/O networking and connectivity demand drivers and challenges include:

- Access to general Web servers and email
- Impact on faster servers having to wait for slower I/Os to complete
- IT services users access to data center applications and servers
- IP-based telephone and other communications services
- Data sharing and movement between facilities and remote offices
- Remote backup and restore of data using managed service providers
- Enabling workload and application load balancing and resource sharing
- Support collaborative and other Web 2.0 applications
- Local area networking among users, clients, and servers
- Access of direct attached, dedicated, and internal-to-external networked shared storage
- Local, regional, and international business continuity and disaster recovery (BC/DR)
- Facilitating remote workforces from virtual offices and small offices/home offices (SOHOs)
- Enabling access to cloud, managed service, and software as a service (SaaS) solutions
- Shifting workloads to take advantage of energy-saving opportunities

Other demand drivers for I/O and networking capabilities include increased use of WebEx for online conferences and the corresponding increase in use of animation, multimedia graphics, video and audio boosting the size of the content and the corresponding impact on networks. Time-sensitive IP-based telephone services, Web TV, instant messaging (IM) along with text messaging and email also continue to put more pressure on existing networks. Additional demand drivers include the growth of online service providers, including managed service providers for Internet access, email, Web hosting, file sharing, storing digital photos online, data backup, BC/DR, archiving, and other online Web- or cloud-based storage services.

Some example of increased networking performance and bandwidth service requirements for voice and video-on-demand include Internet TV, MP4, and MP2 standard-definition video. High-definition MP2 video requires 18–25 Mbit per second or just over 3 MB per second for uncompressed data, with higher resolution requiring even more performance. Media and entertainment production, including computer-generated animation and special-effects rendering along with other postprocessing, requires large amounts of network bandwidth to connect high-performance servers with storage and artists.

Another driver placing more demand on networking capabilities locally and on a wide area or global basis is the growth of Internet-based data management services. These services range from online email and website hosting to portals for storing vacation videos and photos. Software or SaaS as well as cloud services have also become popular for accessing free or low-cost storage space in which to park photos, videos, files, or other digital data online. Other examples include Web-based managed service providers that provide remote backup capabilities for SOHOs, remote offices/branch offices (ROBOs), and small to medium-size businesses (SMBs) as well as consumers. Some managed service providers have also expanded their backup or storage space for rent (or giveaway) capabilities to support remote replication functionality for BC/DR.

Storage as a service is not a new concept: It was tried during the dot.com era, but with little to some success. What is most promising today is the combination of increased bandwidth at lower cost and improved accessibility. Over the past decade, technology maturation, including management tools and software as well as security, and lower storage and networking costs, has come a long way.

Even more promising is the right focus of storage and cloud-based services on consumers; especially SMBs, SOHOs, and ROBOs, whose data and storage needs have increased significantly over the past decade. In the past these target markets usually fell below the radar—not large enough to really need Web-based managed services let alone able to afford them. With today's lower costs and more robust technologies as well as ease of use and deployment, Web-based cloud-based storage look more promising for some environments. For example, while writing this book and traveling internationally, using 3G wireless cells along with traditional WiFi capabilities, I was able to routinely send completed work to a Web-based parking spot in addition to keeping local copies on various media. Enhanced networking capabilities also have a green or PCFE impact in that as improvements continue to support the virtual office and desktop, people can work from home or satellite offices, reducing time spent in transit and therefore reducing emissions and other transportation impacts.

Network operations control centers (NOCCs) use networks to monitor various facilities, server, storage, and networking resources, taking corrective action as needed while collecting event and usage activity information including performance, error rates, and security-related metrics. Networks and I/O are also important to facilitate timely and efficient IRM activities including server, storage, network configuration, management, monitoring, diagnostics, and repair functions.

Facility management, including monitoring and automated response for smoke detection and fire suppression as well as enabling smart and precision cooling, also rely on networking capabilities. For example, using **Simple Network Management Protocol (SNMP),** management information blocks, and IRM software, servers running IRM tools can monitor and take proactive as well as reactive steps to optimize heating, ventilation, and air conditioning (HVAC) of computer room air conditioning (CRAC). In addition, IRM tools can also monitor and manage servers, storage, and networking devices, taking proactive action in various events such as powering down if power or cooling capabilities are in jeopardy. Another IRM function enabled by networking is data protection and data management, including backup, snapshots, replication, and data movement for routine operations as well as to enable high availability and BC/DR.

9.2   Fundamentals and Components

To address the challenges and demands, networks and I/O need to support good performance, flexibility, and reliability in an affordable manner. Given the diversity of applications and usage needs as well as types of servers and storage devices, there are many different protocols and network transports. Networks and I/O interfaces range in price, performance, and best uses for local and wide area needs.

I/O and networking components include:

- Host bus adapters (HBAs), host channel adapters (HCAs)
- Network interface cards (NICs) and network adapters

- Embedded, blade, stackable switches and directors
- Routers and gateways for protocol conversion, distance, segmentation, and isolation
- Specialized appliances for security, performance, and application optimization
- Distance and bandwidth optimization for remote data access and movement
- Managed service providers and bandwidth service providers
- Diagnostic and monitoring tools including analyzers and sniffers
- Cabinets, racks, cabling and cable management, and optical transceivers
- Management software tools and drivers

Block-based storage access involves a server requesting data or writing data to a storage device, specifying a start and stop or range of bytes to be processed, usually in increments of blocks of storage. Storage is organized into blocks of 512-, 1,024-, 2,048-byte or larger blocks. Another form of data access is file-based; instead of requesting a range of blocks to be retrieved or stored, data is accessed by requesting and accessing a file. Ultimately, all data written to storage is handled via block-based access, which is the underlying foundation for storing data.

Most applications and users, however, interact with a file system or volume manager of some type that operates on a file basis. File requests are resolved and processed as blocks by a file system and volume manger or by a file server or with block attached storage. Common file-based access protocols include Network File System (NFS), Parallel NFS (pNFS), and Windows Common Internet Format (CIFS), also known as SAMBA. Object- and message-based access of information or services are also commonly used for accessing websites, email, or other applications. Figure 9.2 shows various networking and storage I/O interfaces and protocols to support different application needs.

Fibre Channel SAN Layers	Description	OSI Layer	IP Routed Network	GbE Network
File system, FTP, SCSI-3, NFS,	Application Presentation Session	7 6 5	FTP, Telnet, HTTP iSCSI, NFS	FTP, Telnet, HTTP iSCSI, NFS
FCP, IPFC, FICON				
FC-4 ULP	Transport	4	TCP, UDP	
FC-3 Services	Network	3	IP	TCP (IP & UDP)
FC-2 Framing, Flow Flow Control	Data link	2	LAN, WAN, MAN	MAC Client MAC Client MAC
FC-1 Encoding, Link FC-0 Physical	Physical	1	Physical	Physical

Figure 9.2   Positioning Various I/O and Networking Technologies and Protocols

9.3   Tiered Access for Servers and Storage—Local and Remote

There is an old saying in the IT industry that the best I/O, whether local or remote, is an I/O that does not have to occur. I/O is an essential activity for computers of all shapes, sizes, and focus in order to be able to read and write data to memory (including external storage) and to communicate with other computers and networking devices (including Internet services). The challenge of I/O is that some form of connectivity, with associated software and time delays, is required while waiting for reads and writes to occur. I/O operations that are closest to the CPU or main processor should be the fastest and occur most frequently for access to main memory using internal local CPU-to-memory interconnects.

I/O and networking connectivity has similar parallel characteristics to memory and storage (see Chapter 7, Figure 7.5). For example, in Figure 9.3, the activity that is closest to the main processor has the fastest I/O connectivity; however, it will also be most expensive, distance-limited, and require special components. Moving away from the main processor, I/O remains fairly fast with distance but is more flexible and cost-effective. An example is the PCI express (PCIe) bus and I/O interconnect, which is slower than processor-to-memory interconnects but is still able to support attachment of various device adapters with very good performance in a

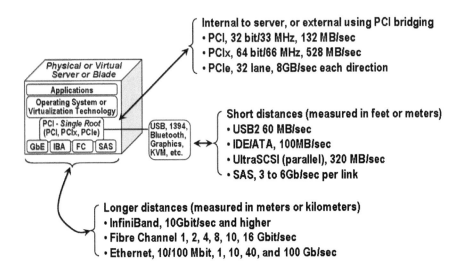

Internal to server, or external using PCI bridging
• PCI, 32 bit/33 MHz, 132 MB/sec
• PCIx, 64 bit/66 MHz, 528 MB/sec
• PCIe, 32 lane, 8GB/sec each direction

Short distances (measured in feet or meters)
• USB2 60 MB/sec
• IDE/ATA, 100MB/sec
• UltraSCSI (parallel), 320 MB/sec
• SAS, 3 to 6Gb/sec per link

Longer distances (measured in meters or kilometers)
• InfiniBand, 10Gbit/sec and higher
• Fibre Channel 1, 2, 4, 8, 10, 16 Gbit/sec
• Ethernet, 10/100 Mbit, 1, 10, 40, and 100 Gb/sec

Figure 9.3   Tiered I/O and Networking Access

cost-effective manner. Farther from the main CPU or processor, various networking and I/O adapters can attach to PCIe, PCIx, or PCI interconnects for backward compatibility to support various distances, speeds, types of devices, and cost factors.

In general, the faster a processor or server is, the more prone to a performance impact it will be when it has to wait for slower I/O operations. Consequently, faster servers need better-performing I/O connectivity and networks. Better performing means lower latency, more IOPS, and improved bandwidth to meet application profiles and types of operations.

9.3.1   Peripheral Component Interconnect (PCI)

Having established that computers need to perform some form of I/O to various devices, at the heart of many I/O and networking connectivity solutions is the **Peripheral Component Interconnect (PCI)** interface. PCI is an industry standard that specifies the chipsets used to communicate between CPUs and memory and the outside world of I/O and networking device peripherals.

Figure 9.4 shows an example of a PCI implementation including various components such as bridges, adapter slots, and adapter types. PCIe leverages multiple serial unidirectional point-to-point links, known as lanes, in contrast to traditional PCI, which used a parallel bus design. In traditional PCI,

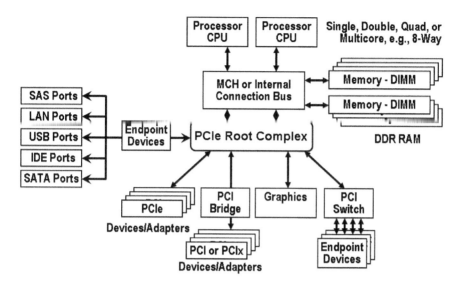

Figure 9.4   PCI Single-Root Configuration Example

bus width varied from 32 to 64 bits; in PCIe, the number of lanes combined with PCIe version and signaling rate determine performance. PCIe interfaces can have 1, 2, 4, 8, 16, or 32 lanes for data movement, depending on card or adapter format and form factor. For example, PCI and PCIx performance can be up to 528 MB per second with a 64-bit, 66-MHz signaling rate, and PCIe is capable of over 4 GB (e.g., 32 Gbit) in each direction using 16 lanes for high-end servers.

The importance of PCIe and its predecessors is a shift from multiple vendors' different proprietary interconnects for attaching peripherals to servers. For the most part, vendors have shifted to supporting PCIe or early generations of PCI in some form, ranging from native internal on laptops and workstations to I/O, networking, and peripheral slots on larger servers.

PCI has become so prevalent that, for example, on a small computer there may not be an external-facing PCI slot for expansion; however, video graphics, USB, Ethernet, serial, WiFi, and other peripheral device ports connect internally via some form of PCI infrastructure. For larger servers, networking, and storage devices, PCI enables adapters or chipsets for networking, storage, and I/O devices. For example, a storage system with 4GFC or 8GFC Fibre Channel, 10-GB Ethernet, SAS, or InfiniBand ports for attachment to servers and networks may have an internal PCI, PCIx, or

PCIe interface between adapter ports or chipsets and the main controller or processing unit of the storage system.

The most current version of PCI, as defined by the PCI Special Interest Group (PCISIG), is PCI Express (PCIe). Backwards compatibility exists by bridging previous generations, including PCIx and PCI, off a native PCIe bus or, in the past, bridging a PCIe bus to a PCIx native implementation. Beyond speed and bus width differences for the various generations and implementations, PCI adapters also are available in several form factors and applications. Some examples of PCI include PCIx and PCIe implementations such as Ethernet, Fibre Channel, Fibre Channel over Ethernet, Infini-Band, SAS, SATA, SCSI, Myrinet, USB, and 1394 Firewire, as well as many specialized functions such as analog-to-digital data acquisition, video surveillance, or other data collection and import needs.

Traditional PCI was generally limited to a main processor or was internal to a single computer, but current generations of PCIe include support for PCI SIG I/O virtualization (IOV), enabling the PCI bus to be extended to distances of a few feet. Compared to local area networking, storage interconnects, and other I/O connectivity technologies, a few feet is very short distance, but compared to the previous limit of a few inches, extended PCIe provides the ability for improved sharing of I/O and networking interconnects.

9.3.2   Local Area Networking, Storage, and Peripheral I/O

A best practice in general is to apply the right tool and technique to the task at hand. Given the diversity of demands and needs that computers serve, there are many different types of technologies to address various needs. For example, there are interfaces and protocols for storage access and general networking as well as peripheral attachments including printers, keyboards, cameras, and other devices. Similar to how PCI has evolved to become the ubiquitous interconnect for attaching various networking and I/O adapters or connectivity devices to computers, consolidation and convergence is also occurring externally from computers and storage devices. For example, networks have converged from early standard solutions such as FDDI, ATM and Token Ring as well as proprietary vendor networks to Ethernet and its many derivatives. Similarly, networking protocols have evolved from vendor-specific implementations to industry-standard TCP/IP or TCP/UDP along with associated application protocols.

Storage and I/O interconnects have also evolved from various proprietary interfaces and protocols to industry-standard Fibre Channel, InfiniBand, Serial Attached SCSI (SAS), and Serial ATA (SATA), as well as Ethernet-based storage. With the exception of IBM legacy mainframes that utilize count key data (CKD) or extended count key data (ECKD) formats and protocols, open system computers, networking, and storage devices have standardized on the SCSI command set for block I/O. Parallel SCSI cabling still exists but is giving way to SAS, SATA, Fibre Channel, iSCSI, and NAS solutions. The SCSI command set, however, continues to exist.

Another example of I/O and networking convergence is the USB interface commonly used for attaching cameras, PDAs, webcams, cell phones, FLASH memory sticks, or USB thumb drives to computers. USB is a short-distance (measured in feet), low-cost, medium-performance (1.5–60 MB per second) interconnect for attaching general-purpose peripherals. USB is also used for attaching serial devices such as dial-up modems, keyboards, printers, and other peripheral devices.

The plethora of protocols and network will continue to migrate toward convergence. In the meantime, multiple interfaces and protocols are used to enable tiered access of data and information. Tiered access enables the most applicable tool to be used for the given task, factoring cost, performance, availability, coexistence, and functionality with application service needs. This includes Fibre Channel at different speeds, iSCSI, InfiniBand, NAS, SAS, and other ways to align the access to the level of service needed. Figure 9.5 shows an example of how different storage and I/O networking protocols and interfaces coexist to meet various application requirements.

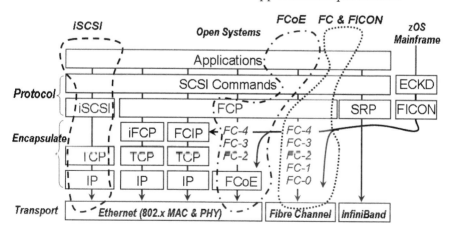

Figure 9.5   Positioning of Data Center I/O Protocols, Interfaces, and Transports

Looking at Figure 9.5, a common question is why so many different networks and transports are needed—why not just move everything to Ethernet and TCP/IP? Converged network architecture will be discussed later in this chapter, showing how, in fact, the number and type of protocols and interfaces continue to converge. For now, the simple answer is that the different interfaces and transports are used to meet different needs, enabling the most applicable tool or technology to be used for the task at hand.

In general, the number of protocols has essentially converged to one for open system block I/O—the SCSI command set or its derivatives—while IBM zOS mainframes and sibling operating systems still use ECKD and FICON, or ESCON in some cases, for I/O and storage protocols. Looking at Figure 9.5, convergence is shown with open system SCSI commands being mapped to Fibre Channel coexisting with IBM mainframe FICON on the same Fibre Channel transport interface. Also shown in Figure 9.5 is the evolving next step in convergence, with Fibre Channel and its upper-level protocols (SCSI_FCP and FC-SB2 FICON) encapsulated and mapped onto an enhanced Ethernet with low latency and lossless characteristics.

9.3.3   Ethernet

Ethernet is a popular industry-standard networking interface and transport medium. Over the past several decades, Ethernet has replaced various proprietary networking schemes, boosting speeds from 10 Mbit to 10 Gbit per second, with 40 and 100 Gbit per second in development. Ethernet is deployed on different physical media, including fiber optic and copper electrical cabling. Versions of Ethernet include 10 Mbit, 100 Mbit (fast), 1,000 Mbit (1GbE), and 10,000 Mbit per second (10 GbE), and emerging 40 and 100 GbE. The new 10-Gbit, 40-Gbit, and 100-Gbit Ethernet standards can be used as an alternative to SONET/SDH for metropolitan area networking environments, leveraging dedicated dark fiber optic cables that are not currently being used to support a network. To further expand the capability and capacity of a fiber optic cable, Wave Division Multiplexing (WDM) and Dense Wave Division Multiplexing (DWDM) can be used to create separate virtual optical data paths, providing additional bandwidth over different optical wavelengths, similar to the colors seen in a rainbow. Normally, with optical networking, a single light color or wavelength is used for transmitting data.

However, light can be divided into multiple subwavelengths known as lambdas with which, using WDM and DWDM technology, different wavelengths can be used simultaneously for sending and receiving data in parallel. For example, a fiber optic cable dedicated to a 10-Gbit Ethernet network can only provide 10Gbits of performance. However, if an 8-lambda WDM or DWDM device is attached to the two ends of the fiber optic cable, the effective performance of the cable between the multiplexing devices (DWDM and WDM) could be 80 Gbits per second. If you are interested in learning more about local, metropolitan, and wide area networking and about optical networking with WDM and DWDM, check out my book, *Designing Resilient Storage Networks*.

In addition to various speeds, cabling media, and topologies, Ethernet encompasses several other capabilities, including link aggregation, flow control, quality of service, virtual LAN (VLAN), and security (shown in Figure 9.6). Other capabilities include support for power over Ethernet to provide electrical power to low-powered devices, simplifying cabling and management. No longer associated with just LAN access of computers and applications, Ethernet is also being used for metropolitan and wide area services along with supporting storage applications.

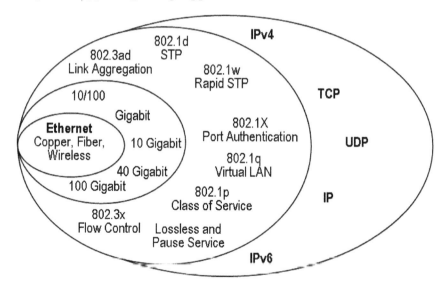

Figure 9.6   Ethernet Ecosystem and Features

9.3.4   Fibre Channel: 1GFC, 2GFC, 4GFC, 8GFC, 16GFC

Fibre Channel is a popular underlying I/O connectivity technology that supports multiple concurrent upper-level protocols (ULPs) for open systems and mainframe server-to-storage, storage-to-storage, and, in some cases, server-to-server I/O operations. Fibre Channel ULPs include FC-SB2, more commonly known as FICON, along with SCSI Fibre Channel Protocol (aka FCP), which is commonly referred to simply as Fibre Channel. Fibre Channel has evolved from propriety implementations operating at under 1Gbit per second to shared-loop 1 Gbit (1GFC) to 1GFC and 2GFC switched, and, more recently, 4GFC and 8GFC. 10GFC has been used mainly for trunks and interswitch links (ISLs) to support scaling and building backbone networks between switches. For the future, 16GFC is on the Fibre Channel Industry Association (FCIA) roadmap, as are other longer-term enhancements.

Most Fibre Channel deployments utilize fiber optics, with electrical connections used mainly in the backplanes of servers, storage, and networking devices. Fibre Channel distances can range from a few meters to over 100 km depending on distance-enabling capabilities such as optics and flow control buffers as well as adapters, switches, and cabling. With the emergence of Fibre Channel over Ethernet (FCoE), utilizing a new and enhanced Ethernet, much of the existing Fibre Channel base can be expected to migrate to FCoE, with some switching to iSCSI, NAS, Infini-Band, or staying on dedicated Fibre Channel.

If you are currently using Fibre Channel for open systems or FICON for mainframe or mixed mode (Fibre Channel and FICON concurrently) and have no near-term plans to migrate open systems storage to IP-based storage using iSCSI or NAS, then you should consider FCoE in addition to near-term 8-Gbit Fibre Channel (8GFC).

9.3.5   Fibre over Ethernet (FCoE)

Ethernet is a popular option for general-purpose networking. Moving forward, with extensions to support FCoE with enhanced low-latency and lossless data transmission, Ethernet will eliminate the need to stack storage I/O activity onto IP. IP will remain a good solution for spanning distance or using for NAS, or as a low-cost iSCSI block-based access option coexisting on the same Ethernet. Getting Fibre Channel mapped onto a common

Ethernet converged or unified network is a compromise among different storage and networking interfaces, commodity networks, experience, skill sets, and performance or deterministic behavior.

For the foreseeable future, FCoE will remain for local environments and not for long-distance use. Unlike iSCSI, which maps the SCSI command set onto TCP/IP, or FCIP, which maps Fibre Channel and its ULPs onto TCP/IP for long distance data transmission to enable remote replication or remote backup, FCoE runs native on Ethernet without the need to run on top of TCP/IP for lower latency in a data center environment.

For long-distance scenarios such as enabling Fibre Channel or FICON remote mirroring, replication, or backups to support high availability or BC/DR, or clustering requiring low-latency communications, use FCIP, DWDM, SONET/SDH, or time-division multiplexing (TDM) MAN and WAN networking solutions and services. For IP-based networks, DWDM, SONET/SDH, Metro Ethernet, and IPoDWDM can be used.

FCoE is very much in its infancy, existing for all practical purposes as a technology demonstration or vendor qualification vehicle. It will be sometime in 2009 or early 2010 before FCoE is ready for mission-critical primetime mass adoption in enterprise environments outside of early-adopter or corner-cases scenarios. With FCoE, the option for a converged network will exist. The degree of convergence and the path to get there will depend on timing, preferences, budget, and other criteria as well as vendor storage offerings and support. As with other techniques and technologies, the applicable solution should be aligned to meet particular needs and address specific pain points while not introducing additional complexity.

Given that FCoE will require a different, more expensive, converged enhanced Ethernet (CEE), iSCSI can continue to leverage the low-cost economic value proposition that enables it to expand its footprint. For existing open systems and IBM mainframe environments that are using Fibre Channel and FICON, the next upgrade option is to go from 4GFC to 8GFC and then, in three years or so, reassess where 16GFC is at along with the status of FCoE ecosystem. For open system environments that are heavily invested in Fibre Channel, the natural progression will be from 4GFC to 8GFC, with some attrition due to shifting over to iSCSI and NAS for some applications.

For environments that are not as heavily invested or committed to Fibre Channel, the opportunity to jump to 10GbE iSCSI will be appealing for some. For those who do make the commitment to at least one more round of Fibre Channel at 8GB, in three to four years it will be time to decide whether to stay with legacy Fibre Channel, assuming 16GFC is ready, jump to FCoE at 10 Gbit or the emerging 40 Gbit, jump to iSCSI, or use some combination.

9.3.6   InfiniBand (IBA)

InfiniBand (IBA) is a unified interconnect that can be used for storage and networking I/O as well as interprocess communications. IBA can be used to connect servers to storage devices, storage to LANs, and servers to servers, primarily for applications within the data center. As a unified interconnect, IBA can be used as a single adapter capable of functioning as multiple logical adapters. IBA enables a channel to extend outside of a single server, up to about 100 m (greater distances may be possible with future iterations), whereas current bus distances are measured in inches. IBA enables memory-to-memory (DMA) transfers to occur with lower overhead, improving storage, networking, and other activities.

A host adapter for IBA, called a host channel adapter (HCA), connects a server to a switch for attachment to storage devices or other servers. Protocols that are supported on IBA include TCP/IP for NAS, iSCSI, as well as support for server-to-server communications including clusters. SCSI Remote Protocol (SRP) maps the SCSI command set for block storage access onto IBA. Gateways and others enable IBA networks to communicate with Ethernet and Fibre Channel networks and devices. IBA has found success in high-performance compute or extreme compute scaling environments, where large numbers of servers require high-performance, low-latency communication, including for cluster and grid applications.

Another use for IBA has been to combine the physical networking and HCAs with a gateway or router attached to Fibre Channel and Ethernet networks with software to create virtual and converged network adapters. For example, an InfiniBand HCA is installed into a server and, with software and firmware, the operating system or virtual machine infrastructure sees what appears to be a Fibre Channel adapter and Ethernet NIC. The benefit is that for nonredundant configurations, a single physical HCA replaces two separate adapters, one for Fibre Channel and one for Ethernet. This is useful where servers are being virtualized to reduce power, cooling, and floor space needs as well as for environments where redundancy is needed yet adapter slots are constrained. Additional coverage of converged networks and I/O virtualization, including Fibre Channel over Ethernet and Converged Enhanced Ethernet (CEE), takes place.

9.3.7   Serial Attached SCSI (SAS)

Serial attached SCSI (SAS) as a technology has several meanings. SAS can and is being used today as a means of attaching storage devices to servers as well as being a means of attaching hard disk drives (HDDs) to storage systems and their controllers. The SAS ecosystem is broad, ranging from protocol interface chips, interposers, and multiplexers to enable dual port of SATA disks to SAS interfaces, SAS expanders, SAS host-based PCI-X and PCIe, as well as external RAID controllers and 3.5-inch and 2.5-inch SAS HDDs, among other components. SAS HDDs are also being deployed inside storage systems (See Figure 9.7) across different price bands and market segments for both block and NAS file storage solutions. SAS product offerings vary depending on their target market or specific solution target price band for deployment with components first, followed by adapters and entry-level block and NAS storage systems, followed by enterprise-class solutions.

SAS HDDs are either replacing or serving as an alternative for Fibre Channel, parallel SCSI, and in some cases serial ATA (SATA) disks, or coex-

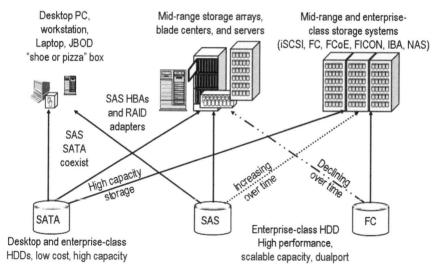

Figure 9.7   Various Uses and Locations of SAS as a technology

isting with SATA devices. In addition to being used to attach HDDs to servers and storage systems controllers, SAS is also being used to attach storage systems—for example, mid-range and entry-level storage arrays—to servers as an alternative to Fibre Channel, parallel SCSI, or iSCSI for open system block storage access.

Common SAS storage technology deployment scenarios include dedicated direct attached storage—DAS internal with SAS HDDs on a server blade, SAS HDDs in a traditional server for boot or local data replacing parallel SCSI or SATA HDDs, dedicated HDDs in a NAS appliance either as primary and expansion storage, or as primary storage with expansion using FC or SATA-based HDDs. Another type of deployment is dedicated direct attached storage—DAS external with SAS HDDs or a mix of SAS and SATA HDDs configured in a storage enclosure (JBOD) or RAID array attached via SAS to a single server using a SAS host controller or SAS host RAID adapter. The storage may be served via NFS or CIFS to other servers and clients to support NAS. Yet another deployment scenario is shared DAS—external with SAS HDDs on a storage blade in a blade center, SAS HDDs or a mix of SAS and SATA HDDs in a storage or RAID array or storage appliance that attaches to two or more servers via SAS.

Another scenario is shared SAN or NAS external storage with SAS HDDs attached to servers via iSCSI, Fibre Channel, or even InfiniBand (SRP) for block access and NAS (NFS or CIFS) over Ethernet or Infini-Band IP-based networks for file-based data and storage-sharing access. Some storage systems support multiple concurrent storage protocols for attachment to servers that can include some combination of iSCSI or NAS using Gb Ethernet, SRP or NAS via IP on InfiniBand, SAS and Fibre Channel including FICON, as well as FCP (SCSI). Similarly, some storage systems support SAS as a primary HDD in the base storage system, with some vendors supporting SAS expansion modules while others support FC and FC-SATA expansion modules. For example, a storage system that has SAS HDDs in the main storage system may not necessarily have SAS expansion drives.

Attributes that make SAS appealing as both a server-to-storage and storage controller-to-HDD interconnect for DAS (SAS), SAN (Fibre Channel and iSCSI), NAS, and storage appliances such as virtual tape libraries or de-duplication appliances include:

- Availability, with native dual port to attach to multiple servers or controllers

- Performance, including 3 and 6 Gbit per second, with 12 Gbit per second on the horizon

- Capacity to combine four links, aggregating port bandwidth,

- Support for more devices and performance over longer distances than with parallel SCSI

- Flexibility, for attachment of SAS and SATA devices at the same time

- Energy-saving opportunities with new HDDs as well as energy-efficient SAS chipsets

- Flexibility for high-performance SAS and low-cost, high-capacity SATA to coexist

9.3.8   Serial ATA (SATA)

Serial ATA (SATA) has gained wide attention over the past several years as a replacement for the parallel ATA/IDE interfaces found on older servers or computers. The popularity of SATA derived from the adoption of low-cost, high-capacity parallel ATA (PATA) disk drives traditionally found in volume servers or personal computers for use in high-density enterprise archiving, backup, and low-cost storage applications. With the shift from ATA to SATA as a storage medium, lower-cost, desktop, and volume storage devices are transitioning to the SATA interface.

SATA-based storage devices continue to be popular given their high capacity of over 1.5 TB per disk drive and relatively low cost to meet bulk storage, near-line, offline, and other applications where higher-performing enterprise-class disk drives are not needed. SATA is also being used as a means of attaching FLASH SSD devices to servers, workstations, and laptops as a low-power alterative to traditional rotating magnetic disk drives. SATA devices can attach to a dedicated SATA controller, adapter, or chipset or to a SAS-based controller to coexist and share a common connection and controller infrastructure.

9.3.9   TCP/IP

Transmission Control Protocol (TCP) and **Internet Protocol (IP)** are core protocols for supporting local and wide area application access and data movement. If there were such a thing as a virtual network, it would be TCP/IP, at least from a nonphysical network transport perspective. The reason that TCP/IP can be considered a virtual network is the frequency with which it is deployed along with its transparency to run over various local and wide area networks to meet various price and performance requirements. TCP/IP is in a transition state, evolving from IPv4 to IPv6 and shifting from a 32-bit address to 128 bits, to boost the number of addresses available in a network

To boost performance and offload servers from having to perform TCP/IP operations, TCP offload engine (ToE) adapters or NICs are available. The idea with ToEs is to offload the processing of TCP networking operations to a specialized card instead of using server CPU resources. While ToEs are valuable for offloading compute overhead from servers to support fast networks, the added cost of specialized ToE adapters, relative performance improvements by servers, and faster Ethernet networks have stymied ToE adoption for all but high-performance servers or storage devices.

TCP/IP is commonly deployed on network transports including WiFi, Ethernet, and InfiniBand, along with wide area optical networks. TCP/IP is used today for both general access of applications as well as for server-to-server communication including cluster heartbeat and data movement. TCP/IP is also being used increasingly for data storage access and movement, including iSCSI block access, NAS for file access and data sharing, as well as for remote mirroring and replication using FCIP.

Popular applications and protocols that leverage TCP/IP include:

- HTTP—HyperText Transfer Protocol for serving and accessing Web pages
- DHCP—Dynamic Host Configuration Protocol for managing network connectivity
- FCIP—Fibre Channel on IP for long-distance data access, movement, and mirroring
- iSCSI—SCSI command set mapped to IP

- NFS—Network File System (NAS) for file and data sharing
- CIFS—Common Internet File System (NAS) for file and data sharing
- pNFS—Parallel NFS (NAS) for parallel high-performance file access
- RTSP—Real-Time Streaming Protocol for streaming data
- MAPI, POP3, SMTP—Microsoft Exchange access
- FTP—File Transfer Protocol
- DNS—Domain Name System for managing Internet domain names
- SNMP—Simple Network Management Protocol for monitoring and device management
- RPC—Remote Procedure Call for program-to-program communications

9.4   Abstracting Distance for Virtual Data Centers

Distance enabling networking and I/O technology is important for virtual data centers to allow data to be moved among locations, remote users, and clients to access data and support traveling workers or home-based workers. Wide area networks need to be safe and secure for access by users of IT resources, including virtual and physical private networks. With the growing popularity of cloud, SaaS, and managed service solutions, wide area networks take on an additional role in (see Figure 9.8).

Virtual and physical data centers rely on various wide area networking technologies to enable access to applications and data movement over distance. Technologies and bandwidth services to support wide area data and application access include DSL, DWDM, Metro Ethernet Microwave, MPLS, Satellite, SONET/SDH, IPoverDWDM, 3G, WiFi and WiMax, T1, T3, optical networking using OC3, OC12, and OC48, as well as other wavelength and bandwidth services.

9.4.1   Metropolitan and Wide Area Networks

Distance is both a friend and a foe to networking. Distance enables survivability and continued access to data. The downside is the cost penalty in

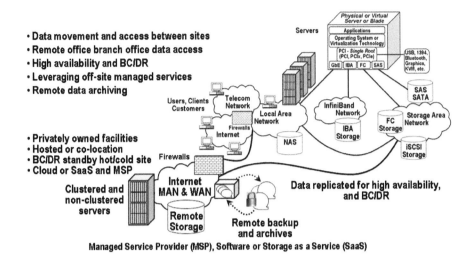

Figure 9.8   Wide Area and Internet Networking

terms of expense, performance (bandwidth and latency), and increased complexity. In looking at networks to span distances, bandwidth is important but latency is critical for timely data movement to ensure data consistency and coherency.

Some examples of how metropolitan (MAN) and wide area (WAN) networks are used include:

- Remote and hosted backup and restore
- Data archiving to managed or hosted remote location
- Data snapshots and replication for high availability and BC/DR
- Data movement and migration plus distribution services
- Remote access of centralized or distributed resources and applications

High-speed WANs are becoming indispensable for critical applications. Business continuance, remote mirroring and replication, cloud computing, SaaS, access to MSPs, and connecting regional data centers are all tasks that require optical WANs. With many optical options and prices, picking the best technologies and techniques to transport data is challenging.

Most storage applications are time-sensitive and require high throughput (bandwidth) and low latency with zero data loss. Bandwidth is the measure of how much data can be transferred over a network or I/O interface in a particular time, for example, per second. Latency, also known as response time, is how long it takes for an I/O activity or event to take place or how much delay occurs in sending or receiving data. Effective bandwidth is a measure of how much of the available bandwidth can actually be used, taking into consideration dropped packets and retransmission due to congestion and protocol inefficiency. A common mistake is to look at bandwidth simply in terms of dollars per gigabit per second. The effective or actual usage is important, and for bandwidth that includes what level of utilization at a given response time (latency level) can be maintained without congestion and packet delay or loss.

A common mistake is to prototype a storage application at a reduced workload and assume that heavier workloads will scale linearly with regard to bandwidth and latency. On the contrary, effective bandwidth can drop off as workload is added. This, along with additional latency, can result in poor performance, particularly for synchronous storage applications. First and foremost, an understanding of the organization's particular needs and goals as well as the capabilities of the various technologies is necessary.

Diverse network paths are important for resilient data and information networks. Bandwidth service providers can provide information as to how diverse a network path exists, including through partner or subcontractor networks. It is also important to determine how the service provider will manage and guarantee network performance (low latency and effective bandwidth). Generally speaking, the more technology layers, including networking protocols, interfaces, devices, and software stacks for performing I/O and networking functions there are, the lower the performance and the higher the latency.

When comparing network and bandwidth services, it is important to look beyond stated line rate or spec sheet numbers, instead learning what the effective performance and corresponding latency will be. For example, a network service provider may offer a lower rate for a shared high speed bandwidth service; however, that may be based on upload or download speeds and be independent of latency. Another provider might offer a higher-cost service that may appear to have the same amount of bandwidth but on closer investigation offers a better effective (usable) bandwidth with lower latency and higher availability.

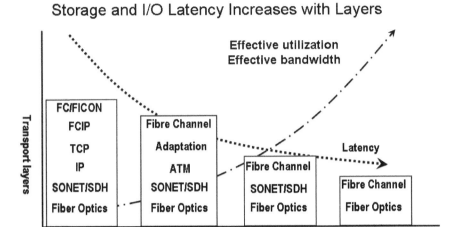

Figure 9.9   Latency and Impact of Adding Technology Layers

Understanding what layers are involved in the network is important because each layer adds complexity, cost, and latency (see Figure 9.9). Latency is important to understand and to minimize for storage and data movement where consistency and transaction integrity are important for real-time data movement. The specific amount of latency will vary from negligible—almost nonexistent—to noticeable, depending on the types of technologies involved and their implementations. In Figure 9.9, using Fibre Channel over a WAN with a combination of technologies including mapping Fibre Channel and/or FICON traffic onto an IP network using FCIP necessitates an additional technology and protocol layer. Depending on how the IP traffic is moved over a network, additional layers and technologies can be involved, for example, IP mapped to SONET, MPLS, IPoDWDM, or other services. The trade-off in performance and latency is the ability to span greater distances using variable-cost networking and bandwidth services to meet specific business requirement or application service-level objectives.

Distance is often assumed to be the enemy of synchronous or real-time data movement, particularly since latency increases with distance. However, latency is the real enemy, because even over short distances, if high latency or congestion exists, synchronous data transmissions can be negatively impacted.

Optical networks are assumed to be and can be faster than electrical or copper wire networks, but the reality can also be very different. While data is able to travel at the speed of light on an optical network, actual performance is determined by how the network is configured and used. For example, in dense wave division multiplexing (DWDM), 32 different networks, each at 40 Gbits per second, are multiplexed onto a single fiber optic cable. The effective performance for a fiber optic cable on its own, in optimal conditions, over a given distance is about 1,280 Gbits per second. However, actual performance is determined by the speed at which the networks connect to each end of the multiplex network. Simply put, the speed of the network is that of the slowest link in the chain.

General considerations pertaining to wide area networking include:

- Distance—how far away do applications and data need to be located?
- Bandwidth—how much data can be moved, and in what timeframe?
- Latency—what are application performance and response time requirements?
- Security—what level of protection is required for data in flight or remote access?
- Availability—what are the uptime commitments for the given level of service?
- Cost—what will the service or capability cost initially and over time?
- Management—who and how will the network service be managed?
- Type of service—dedicated or shared optic or other form of bandwidth service?

Wide area networking options and technologies include:

- Dedicated and dark fiber optic cabling and wavelength services
- Wave division multiplexing (WDM) and dense WDM (DWDM)

- IP over DWDM (IPoDWDM) for IP-based services and metropolitan Ethernet
- SONET/SDH optical carrier (OC) networking and packet over SONET (POS)
- TCP/IP services, networks, and protocols, including FCIP, iFCP, and iSCSI
- Multi-Protocol Label Switching (MPLS) services
- Wireless WiFi, Fixed WiFi, WiMax, and 3G cell phone services
- Satellite and microwave wireless transmission
- Data replication optimization (DRO) and bandwidth optimization solutions
- Firewalls and security applications

9.4.2 Wide Area File Service (WAFS) and Wide Area Application Service (WAAS)

Wide Area File Service (WAFS), also known by vendor marketing names such as wide area data management (WADM) and wide area application services (WAAS), is generically a collection of various services and functions to help accelerate and improve access of centralized data. Whereas data replication optimization accelerates performance and maximizes bandwidth for movement of data to support replication, mirroring, and remote tape backup, WAFS solutions focus on improving the productivity of users accessing centralized data from remote offices. Consequently, for environments that are looking to consolidate servers and storage resources away from ROBO locations, WAFS can be an enabling technology and coexist in hybrid environments to enhance backup of distributed data. DRO technologies, on the other hand, complement remote data replication from NAS and traditional storage along with disk-to-disk-to-tape remote backup.

Wide area file and data services vary in the specific features and applications that are supported. Most, if not all, provide some form of bandwidth or protocol optimization, compression, and other techniques to speed access and movement of data between locations. What differs between these technologies and traditional bandwidth and DRO solutions for storage and data replication is that these new wide area tools are application and data type aware. For example, bandwidth optimizers and compression tools generally brute-force compress and reduce the data footprint of data being sent or accessed over a network. Application, file, and data services are optimized to support specific networking protocols as well as applications—for example, Oracle, SQL, Microsoft Exchange, and AutoCAD, among others—to boost performance without compromising security or data integrity.

9.5   Virtual I/O and I/O Virtualization

Virtual I/O (VIO) and I/O virtualization (IOV) sound similar, but there are distinct differences between them. For example, VIO includes technologies that mask or minimize the impact of performing an I/O operating using RAM or FLASH-based memory, including solid-state disk devices or caching appliances. The aim of VIO is to provide abstraction and transparency to applications without having to perform and wait for I/O operations to complete. IOV, on the other hand, involves emulation and consolidation to improve utilization of I/O adapters and supporting infrastructures, including consolation. Another capability of IOV, also known as converged network solutions, is to simplify and remove the complexity of having to support multiple adapters and associated software.

On a traditional physical server, the operating system sees one or more instances of Fibre Channel and Ethernet adapters even if only a single physical adapter, such as an InfiniBand HCA, is installed in a PCI or PCIe slot. In the case of a virtualized server—for example, VMware ESX—the hypervisor will be able to see and share a single physical adapter, or multiple adapters, for redundancy and performance to guest operating systems. The guest systems see what appears to be a standard Fibre Channel and Ethernet adapter or NIC using standard plug-and-play drivers.

Virtual HBA or virtual network interface cards (NICs) and switches are, as their names imply, virtual representations of a physical HBA or NIC, similar to how a virtual machine emulates a physical machine with a virtual server. With a virtual HBA or NIC, physical NIC resources are carved up and allocated as virtual machines, but instead of hosting a guest operating system like Windows, UNIX, or Linux, a Fibre Channel HBA or Ethernet NIC is presented. Are IOV or VOI a server topic, a network topic, or a storage topic? Like server virtualization, IOV involves servers, storage, network, operating system, and other infrastructure resource management areas and disciplines. The business and technology value proposition or benefits of converged I/O networks and virtual I/O are similar to those for server and storage virtualization.

Benefits and value propositions for IOV include:

- Doing more with what resources (people and technology) already exist or reduce costs
- Single (or pair for high availability) interconnect for networking and storage I/O
- Reduction of power, cooling, floor space, and other "green-friendly" benefits
- Simplified cabling and reduced complexity for server network and storage interconnects
- Boosting server performance to maximize PCI or mezzanine slots
- Rapid re-deployment to meet changing workload and I/O profiles of virtual servers
- Scaling I/O capacity to meet high-performance and clustered application needs
- Leveraging common cabling infrastructure and physical networking facilities

In addition to virtual or software-based NICs, adapters, and switches found in server virtualization implementations, virtual LAN (VLAN), virtual SAN (VSAN), and virtual private network (VPN) are tools for providing abstraction and isolation or segmentation of physical resources. Using emulation and abstraction capabilities, various segments or subnetworks can be physically connected yet logically isolated for management, performance, and security purposes. Some form of routing or gateway functionality enables various network segments or virtual networks to communicate with each other when appropriate security is met.

Additional forms of IOV and VOI include:

- N_Port_ID Virtualization
- Blade center virtual connectivity
- Converged networks

- PCI SIG IOV
- Converged Enhanced Ethernet and FCoE
- InfiniBand converged networks

9.5.1   N_Port_ID Virtualization

N_Port_ID Virtualization (NPIV), shown in Figure 9.10, uses an ANSI T11 Fibre Channel standard to enable a physical HBA and switch to support multiple logical World Wide Node Names (WWNN) and World Wide Port Names (WWPN) per adapter for shared access purposes. Fibre Channel adapters can be shared in virtual server environments across the various virtual machines (VMs), but the various VMs share a common World Wide Node Name (WWNN) and World Wide Port Name (WWPN) address of the physical HBA. The issue with a shared WWNN and WWPN across multiple VMs is that, from a data security and integrity perspective, volume or logical unit number (LUN) mapping and masking have to be performed on a coarse basis.

By using NPIV supported by a target operating system or virtual server environment along with associated HBAs and switches, fine-grained allocation and addressing can be performed. With NPIV, each VM is assigned a unique WWNN and WWPN that is independent of the underlying physical HBA. By having a unique WWNN and WWPN, VMs can be moved to different physical servers without having to make changes for addressing of different physical HBAs or changes to Fibre Channel zoning on switches. In addition, NPIV enables fine-grained LUN mapping and masking to enable a specific VM or group of VMs to have exclusive access to a particular LUN when using a shared physical HBA.

A by-product of the fine-grained and unique WWPN is that a LUN can be moved and accessed via proxy backup servers, such as VMware VCB, when it is properly mapped and zoned. The benefit is that time-consuming changes to SAN security and zoning for new or changed devices do not have to be made when a VM moves from one physical server and HBA to another (see Figure 9.10).

In Figure 9.10, the left server is a primary server and the right server is a standby server. In this example, the primary server has three VMs that share a physical Fibre Channel adapter named as WWN0. (Note that for illustrative purposes, the formal Fibre Channel WWN addressing scheme is being

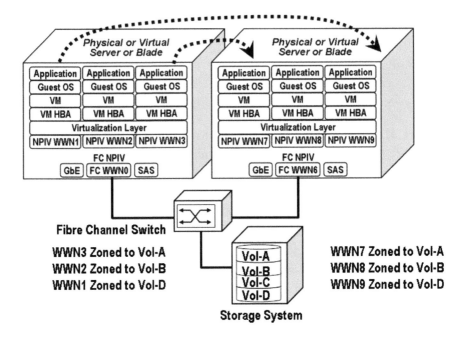

Figure 9.10   Fibre Channel N_Port_ID Virtualization

shortened.) The Fibre Channel adapter in the physical server is configured with NPIV capabilities and each of the three VMs is allocated a unique WWN of WWN1, WWN2, or WWN3. Similarly, the standby server has three VMs, each with a unique WWN of WWN7, WWN8, or WWN9, with the physical Fibre Channel adapter having an example address of WWN6. When the VMs and their guest operating systems and applications are moved from the left primary server to the right standby server, no changes should be required to zoning or LUN or volume mapping configurations, assuming that the Fibre Channel zones and LUN mapping are set up with Vol-A being accessible by WWN3 and WWN7, Vol-B by WWN2, and WWN8 and Vol-D by WWN1 and WWN9.

Benefits and functionality that are enabled by NPIV include:

■ Track and report I/O performance between a particular VM and LUN versus shared ports
■ Fine-grain security using LUN mapping and masking between specific VMs and LUNs

- Transparency for configuration changes without disrupting existing Fibre Channel zones

- WWPN of VM follows the VM when moved independent of physical HBA address

- QoS on a shared physical HBA to enable enhanced performance for specific VMs

- Proxy based backups, including VCB, for LUN tracking and mapping for data protection

- Simplified troubleshooting with insight of individual VM WWPN on shared HBA

- SAN segmentation or routing to a specific VM instead of physical port level

9.5.2   Blade Center and Server Virtual Convexity Features

Blade center chassis that house server blades also contain various I/O and networking connectivity options. Blade servers using mezzanine cards, which are essentially small HBA or NICs that attach to the blade server's primary printed circuit board, support different I/O and networking options. These options include SAS for direct attached shared or dedicated disk drives for operating system boot or local storage. Other networking and I/O connectivity include 1- and 10-Gbit Ethernet, 2-, 4-, and -8Gbit Fibre Channel, and InfiniBand. Embedded switches that are installed in blade center I/O slots communicate with mezzanine cards via blade chassis backplanes to eliminate associated cabling clutter found with external rack-mounted servers.

Blade center switches vary in the number of ports facing in toward the mezzanine cards on the server blades and the number of ports facing outward. Specific vendor switches have different configuration and port options. Usually there are at least as many inward-facing ports as there are blades, with some number of external-facing uplink or trunk ports to attach to core or aggregation switches. Depending on vendor implementation and configuration, the allocation of embedded switch ports to mezzanine cards and outward-facing ports may be on a one-to-one basis or on an oversubscribed basis. For example, if there are 12 server blades and a switch has 14 inward-facing ports, each operating at 4 Gbit in the case of Fibre Channel, and two outward-facing 8-Gbit ports, there will be an allocation ratio of 48

Gbit to 16 Gbit or 3:1 oversubscription. Assuming that not all 12 ports from the servers ever operate at or near 4 Gbit, the oversubscription will not be an issue. However, for higher-performing servers and I/O demands, a lower oversubscription ratio may be needed, such as 2:1 or even 1:1.

For servers that need 1:1 server-to-network I/O performance capabilities, some blade centers and vendors have pass-through ports that enable blades to bypass embedded switches and communicate directly with external switch ports. Blade center vendors have combined software for managing and configuring different types of blades across multiple blade centers and server blades. Management software allows blades to be configured, switch ports allocated, and other routine configuration and management functions to be performed.

9.5.3   Converged Networks

I/O and general-purpose data networks continue to converge to enable simplified management, reduce complexity, and provide increased flexibility of IT resource usage. Converged networks and virtualized I/O are being used at both the server level internally with PCIe enhancements as well as externally with Ethernet, Fibre Channel, and InfiniBand. Even SAS and SATA, discussed earlier in this chapter, are a form of convergence by which SATA devices can attach to a SAS controller and coexist with SAS devices to reduce complexity, cabling, and management costs. Figure 9.11 shows an example using virtual HBAs and NICs attached to a switch or I/O director that in turn connects to Ethernet-based LANs and Fibre Channel SANs for network and storage access. Examples of converged networks include Fibre Channel over Ethernet using an enhanced Ethernet, Fibre Channel and Ethernet virtual HBAs and NICs using InfiniBand as a transport, and inside servers using PCIe IOV.

Figure 9.12 shows how various network and I/O transports and protocols have been converging for several decades. The evolution has been from proprietary vendor interfaces and protocols to open industry standards. For example, SCSI is both a parallel cabling scheme and a protocol command set. The SCSI command set has been implemented on SAS, Fibre Channel, and InfiniBand (SRP), and on IP in the form of iSCSI. LAN networking has evolved from various vendor-specific network protocols and interfaces to standardize around TCP/IP and Ethernet. Even propriety interfaces and protocols such as IBM Mainframe FICON, which evolved from ESCON,

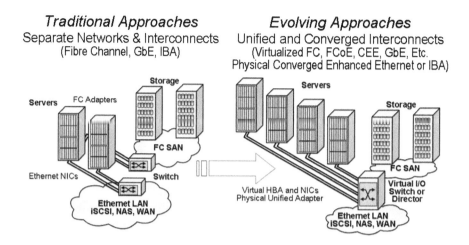

Figure 9.11   Unified or Converged Data Center Fabric or Network

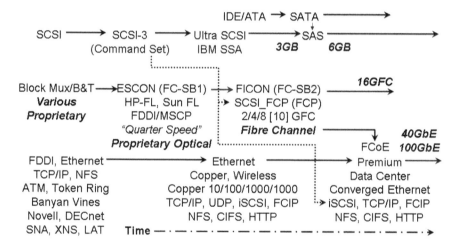

Figure 9.12   Network and I/O Convergence Paths and Trends

on propriety fiber optic cabling, now coexist in protocol intermix mode on Fibre Channel with open systems SCSI_FCP.

Ethernet supports multiple concurrent upper-level protocols (see Figure 9.12), for example TCP/IP and TCP/UDP, along with legacy LAT, XNS, and others, similar to how Fibre Channel supports multiple ULPs including FICON for IBM mainframes and FCP for open systems. Over the past decade, networking and storage I/O interfaces have been on a convergence course, with industry standards providing flexibility, interoperability, and variable cost-to-functionality options.

Propriety mainframe interconnects, such as bus & tag (block mux), gave way to ESCON, an early and propriety derivative implementation of quarter-speed Fibre Chanel (less than 25 MB per second). Later, ESCON gave way to FICON, which leverages common underlying open Fibre Channel components to enable FICON to coexist with open systems Fibre Channel FCP traffic in protocol intermix mode. Similarly, parallel SCSI evolved to UltraSCSI and separation of the SCSI command set from physical parallel electrical copper cables, enabling SCSI on IP (iSCSI), SCSI on Fibre Channel (FCP), SCSI on InfiniBand (SRP), serial attached SCSI (SAS), and other technologies. Traditional networks, including FDDI and Token Ring, have given way to the many different 802.x Ethernet derivatives. Consequently, the continuing convergence evolution is to leverage the lower-level MAC (Media Access Control) capabilities of an enhanced Ethernet. Enhancements around quality of service to improve latency as well as to enable lossless communications allow Fibre Channel and its ULPs to coexist on a peer basis with other Ethernet ULPs, including TCP/IP.

Moving forward, one of the premises of a convergence-enhanced Ethernet supporting FCoE is the ability to carry Fibre Channel traffic, including both FCP and FICON, on an Ethernet that can also transport TCP/IP traffic concurrently. This differs from current approaches, in which Fibre Channel traffic can be mapped onto IP using FCIP for long-distance remote replication. With FCoE, the TCP/IP layer is removed along with any associated latency or overhead, but only for local usage. A common question is why not use iSCSI, why the need for FCoE, why not just use TCP/IP as the converged network? For some environments, where low cost, ease of use, and good performance are the main requirements, iSCSI or NAS access for storage is a good approach. However, for environments that need very low latency, good or very good performance, and additional resiliency, Fibre Channel remains a viable option. For environments that need FICON for mainframes, iSCSI is not an option in that iSCSI is only an SCSI implementation on IP and not a mapping of all traffic.

9.5.4   PCI-SIG IOV

PCI SIG IOV consists of a PCIe bridge attached to a PCI root complex along with an attachment to a separate PCI enclosure. Other components and facilities include address translation service (ATS), single-root IOV

(SR-IOV), and multiroot IOV (MR-IOV). ATS enables performance to be optimized between an I/O device and a server's I/O memory management. Single-root, SR-IOV enables multiple guest operating systems to access a single I/O device simultaneously, without having to rely on a hypervisor for a virtual HBA or NIC. The benefit is that physical adapter cards, located in a physically separate enclosure, can be shared within a single physical server without having to incur any potential I/O overhead via virtualization software infrastructure. MR-IOV is the next step, enabling a PCIe or SR-IOV device to be accessed through a shared PCIe fabric across different physically separated servers and PCIe adapter enclosures. The benefit is increased sharing of physical adapters across multiple servers and operating systems.

Figure 9.13 shows an example of a PCIe switched environment, where two physically separate servers or blade servers attach to an external PCIe enclosure or card cage for attachment to PCIe, PCIx, or PCI devices. Instead of the adapter cards physically plugging into each server, a high-performance short-distance cable connects the server's PCI root complex via a PCIe bridge port to a PCIe bridge port in the enclosure device. In the example, either SR-IOV or MR-IOV can take place, depending on specific PCI firmware, server hardware, operating system, devices, and associated drivers and management software. For a SR-IOV example, each server has access to some number of dedicated adapters in the external card cage, for example, InfiniBand, Fibre

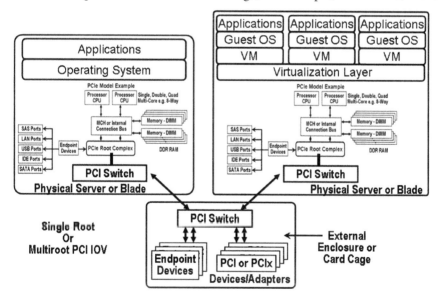

Figure 9.13   PCI SIG IOV

Channel, Ethernet, or FCoE HBAs. SR-IOV implementations do not allow different physical servers to share adapter cards. MR-IOV builds on SR-IOV by enabling multiple physical servers to access and share PCI devices such as HBAs and NICs safely with transparency. The primary benefit of PCI IOV is to improve utilization of PCI devices, including adapters or mezzanine cards, as well as to enable performance and availability for slot-constrained and physical footprint- or form factor-challenged servers. Caveats of PCI IOV are distance limitations and the need for hardware, firmware, operating system, and management software support to enable safe and transparent sharing of PCI devices.

9.5.5   Convergence Enhanced Ethernet and FCoE

Given the broad market adoption of Ethernet for general-purpose networking, including application access, data movement, and other functions, it only makes sense for further convergence to occur around Ethernet. The latest enhancements to Ethernet include improved quality of service and priority groups along with port pause, lossless data transmission, and low-latency data movement. Initially, FCoE and enhanced Ethernets will be premium solutions, but over time, price reductions should be seen, as they have with previous generations of Ethernet.

These and other improvements by the Internet Engineering Task Force (IETF), the organization that oversees Internet and Ethernet standards, as well as improvements by ANSI T11, the group that oversees Fibre Channel standards, have resulted in a converged enhanced Ethernet. Essentially, Fibre Channel over Ethernet combines the best of storage networking with low latency, deterministic performance with lossless data transmission with Ethernet's broad adoption and knowledge base. By mapping Fibre Channel to Ethernet, essentially encapsulating Fibre Channel traffic into Ethernet frames, upper-level protocols of Fibre Channel, including SCSI_FCP and IBM mainframe FICON (FC-SB2), should be able to coexist on the enhanced Ethernet with other Ethernet-based traffic, including TCP/IP.

Figure 9.14 shows a traditionally separate fiber optic cable being dedicated (bottom of figure) in the absence of wave division multiplexing technology. With FCoE, Fibre Channel is mapped onto an enhanced low-latency, lossless with quality of service (QoS) Ethernet to coexist with other traffic and protocols including TCP/IP.

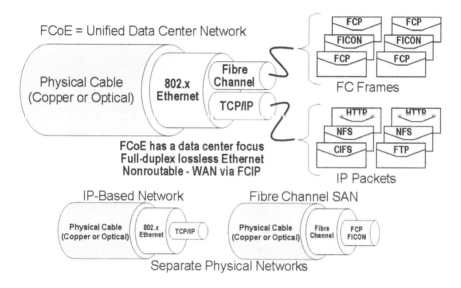

Figure 9.14   Converged Network (Top) and Separate Networks (Bottom)

Note that FCoE is targeted for the data center, as opposed to long distance, and continues to rely on FCIP (Fibre Channel mapped to IP) or a WDM (Wave Division Multiplexing) MAN for shorter distances. For example, in Figure 9.14, the traditional model for cabling a LAN and SAN has separate physical copper or optical cables for each network, unless a DWDM (Dense Wave Division Multiplexing) multiplexed optical network is being used. With FCoE, the next step in the converged network evolution takes place with enhanced Ethernet the common denominator that supports both FCoE and other Ethernet networks concurrently on a single Ethernet network.

Figure 9.15 shows an example with converged NIC and adapters capable of attaching to FCoE storage switches that can also attach to legacy LANs for wide area data movement. Also shown, on the right, is a server attached to a separate converged enhanced Ethernet LAN with other servers, iSCSI, and NAS storage attached. The converged NIC or adapter presents to the servers what appears as a Fibre Channel adapter, Ethernet NIC, or both, depending on the specific vendor implementation and packaging. In the case of a blade server, mezzanine cards provide similar functionality, eliminating the need for separate Fibre Channel and Ethernet cards. With 10GbE and evolving 40GbE technology, sufficient performance should be available for direct to server blade centers as well as high-performance storage devices

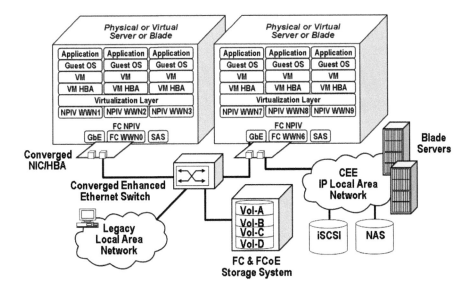

Figure 9.15   FCoE and Converged Enhanced Ethernet

using FCoE and converged Ethernet. For even higher speeds, 100GbE can be used as it matures for backbone and core-to-core networking.

9.5.6   InfiniBand IOV

InfiniBand-based IOV solutions are an alternative to Ethernet-based solutions. Essentially, InfiniBand approaches are similar, if not identical, to converged Ethernet approaches including FCoE, with the difference being InfiniBand as the network transport. InfiniBand HCAs with special firmware are installed into servers that then see a Fibre Channel HBA and Ethernet NIC from a single physical adapter. The InfiniBand HCA also attaches to a switch or director that in turn attaches to Fibre Channel SAN or Ethernet LAN networks.

The value of InfiniBand converged networks are that they exist today, and they can be used for consolidation as well as to boost performance and availability. InfiniBand IOV also provides an alternative for those who do not choose to deploy Ethernet. From a PCFE standpoint, converged networks can be used for consolidation to reduce the total number of adapters and the associated power and cooling. In addition to removing unneeded adapters without loss of functionality, converged networks also free up or allow a reduction in the amount of cabling, which can improve airflow for cooling, resulting in additional energy efficiency.

9.6   Virtualization and Management Tool Topics

Networks and I/O connectivity are important and, with continued convergence, the lines between storage interfaces and networks are becoming blurred. Networks and I/O involve networking, server, storage, and data professionals to coordinate the various interdependencies across the different technology domains. For example, if a virtual server is moved from a VM on one physical server to another VM on a different physical server, unless NPIV is being used, SAN personnel have to make zoning and configuration changes, as do server and storage personnel. Various tasks and activities are required to keep I/O and network capabilities in good order to support virtual data centers.

Common I/O and networking management functions and activities include:

- Global name space management software
- Event correlation and analysis tools
- Troubleshooting and diagnostics
- Provisioning and resource allocation
- Firmware and hardware upgrades
- Asset management and configuration management
- Management security and authorize
- File management tools, including discovery, reporting, and data movement
- Data security tools, including encryption and key management
- Performance and utilization monitoring and reporting

Metrics for networking include:

- Bandwidth of data moved
- Latency and response time

- Frames or packets per second
- Availability and quality of service
- Top talkers between source and destinations
- Error counts including retransmissions

9.6.1   Networking Options for Virtual Environments

What networking and I/O interfaces and protocols to use for a virtual data center, including virtual server and storage scenarios, depends on specific needs and QoS. Other factors include type of I/O profile—large or small, random or sequential reads or writes—and number of VMs and type of device adapters. Depending on the version of a virtual infrastructure, such as VMware or Microsoft, among others, some advanced features are supported only with certain protocols. For example, a VM may be able to boot off of any network interface or protocol, but the virtualization hypervisor may have restrictions or specific configuration requirements.

Although some VMs can use and leverage local dedicated and non-shared storage, most features for scaling and resiliency, including high availability and BC/DR, require some form of shared storage. Shared storage includes dual- or multiported SAS storage arrays attached to two or more servers, iSCSI and Fibre Channel block storage, along with NFS NAS storage. Other features, including clustering, dynamic VM movement or relocation, and server-free backup, vary depending on the version and type of I/O interface and protocol being used.

Performance varies depending on specific VM configuration, underlying hardware architecture, guest operating system and drivers, as well as storage system configuration. Virtualization vendors have configuration guides and online forums covering different configurations and supported options, as do server and storage vendors. Needless to say, there is debate in the storage community among iSCSI, Fibre Channel, and NFS NAS vendors as to which protocol is the most efficient and best to use. Not surprisingly, some NFS vendors claim their protocol and interface as being the best, as do some iSCSI or some Fibre Channel vendors. Others take a more neutral and consultative approach of what is the best tool and technique for the task at hand, considering also individual preferences and existing or planned technology decisions.

Other applications, such as Microsoft Exchange, may requirement running on block-based storage or on a Microsoft-based file system. Traditional databases such as Oracle, Microsoft SQL, or IBM DB2/UDB have had a preference and recommendation, if not requirement, to run on block-based storage. Databases such as Oracle can and do run on NFS storage and, by leveraging a feature in NFS V3 or later called direct I/O (DIO), can perform block-like access of NAS storage without having to read or write an entire file. These rules and requirements are changing, and it is important to consult with vendors or product specialists regarding specific guidelines and recommendations.

9.6.2 Oversubscription: Not Just for Networks

Oversubscription is an indicator of potential performance issues under different workloads, and a design to eliminate oversubscription removes this as a possibility. Congestion is the result of oversubscription when ports have to wait for access to network interfaces and can be blocked from doing their useful work. A minor example of congestion is when you try to dial someone on a telephone and almost immediately get a busy signal. An example of complete blockage would be if you pick up the phone and cannot get a dial-tone even though you know the power is good and the phone lines have not been cut. Telephone networks, like data networks, have some degree of oversubscription built in to help reduce costs. Storage interfaces have traditionally been designed to be nonblocking and nonoversubscribed to eliminate this as a possibility.

Oversubscription results when more ports are present than bandwidth is available to support full-speed communication between the ports. In a single switching device, the core or switching modules should provide adequate bandwidth to be nonblocking or prevent oversubscription by design. This can, however, be overridden on some products that support oversubscribed ports by design, to help drive the cost per port down by sharing switching bandwidth. The design premise is that not all servers will operate and need the full bandwidth, so some servers can attach to ports that will be oversubscribed by design. Approach with caution and an understanding of server I/O workload and profiles to prevent performance challenges from occurring.

9.6.3   Security

Networking enables flexible access of IT resources by support staff, users, and clients on a local and wide area basis. However, with the flexibility, ease of use, and access, security issues also increase as a result of networking with virtual and physical IT resources and applications or services being delivered. For example a non-networked, standalone server and dedicated direct attached storage with secured physical and logical access is more secure than a server attached to a network with general access. However, the standalone server will not have the flexible access of a networked server that is necessary for ease of use. It is this flexible access and ease of use that requires additional security measures.

Examples of various threat risks and issues include:

- Virus, botnets, spyware, root kits, and email spam
- Intrusion by hackers to networks, servers, storage, and applications
- Theft or malicious damage to data and applications
- Lost, misplaced, or stolen data or stolen network bandwidth
- Regulatory compliance and information privacy concerns
- Exposure of information or access to IT resources when using public networks

As with many IT technologies and services, there will be different applicable threat risks or issues to protect against, requiring various tiers and rings of protection. The notion of multiple rings or layers of defense is to allow for flexibility and enable worker productivity while providing protection and security of applications and data. A common belief is that applications, data, and IT resources are safe and secure behind company firewalls. The reality is that if a firewall or internal network is compromised, in the absence of multiple layers of security protection, additional resources will also be compromised. Consequently, to protect against intrusions by external or internal threats, implementation of multiple protection layers, particularly around network access points, is vital.

Networking and I/O security topics and action items include:

- Secure management consoles, software tools, and physical ports on IT technologies.
- Enable intrusion detection and alerts for IT resources including networked devices.
- Check proactively for network leakage including lost bandwidth or device access.
- Physically secure networking devices, cabling, and access points.
- Protect against internal threats as well as external threats.
- Implement encryption of data at rest as well as data in flight over networks.
- Limit access rights to various IT resources while enabling productivity.
- Utilize VLAN and VSAN along with VPN and firewall technologies.
- Implement Fibre Channel SAN zoning, authentication, and authorization.
- Enable physical security in addition to logical security and password management.
- Use multiple layers of security for servers, storage, networks, and applications.
- Use private networks combined with applicable security and defense measures.
- Implement key and digital rights management across applications and IT resources.

9.6.4   Cabling and Cable Management

Wireless networking continues to gain in popularity; however, physical cabling using copper electrical and fiber optic cabling continues to be used. With the increased density of servers, storage, and networking devices, more cabling is being required to fit into a given footprint. To help enable management and configuration of networking and I/O connectivity, networking devices including switches are often integrated or added to server and storage cabinets. For example, a top-of-rack or bottom-of-rack or embedded network switch aggregates the network and I/O connections within a server cabinet to simplify connectivity to an end-of-row or area group of switches. Lower-performing servers or storage can use lower-cost, lower-performance network interfaces to connect to a local switch, and then a higher-speed link or trunk, also known as an uplink, to a core or area switch. Cabling by itself does not require power or cooling, but power can be transmitted over copper electrical cabling for Ethernet applications.

Cable management systems, including patch panels, trunk, and fan-in, fan-out cabling for overhead and under-floor applications, are useful for organizing cabling. Cable management tools include diagnostics to verify signal quality and db loss for optical cabling, cleaning and repair for connectors, as well as asset management and tracking systems. A relatively low-tech cable management system includes physically labeling cable endpoints to track what the cable is being used for. Software for tracking and managing cabling can be as simple as an Excel spreadsheet or as sophisticated as a configuration management database (CMDB) with intelligent fiber optic management systems. An intelligent fiber system includes mechanisms attached to the cabling to facilitate with tracking and identify cabling.

Another component in the taxonomy of server, storage, and networking I/O virtualization is the virtual patch panel that masks the complexity by abstracting the adds, drops, moves, and changes associated with traditional physical patch panels. For large and dynamic environments with complex cabling requirements and the need to secure physical access to cabling interconnects, virtual patch panels are a great complement to IOV switching and virtual adapter technologies.

In addition to utilizing cabling that is environmentally friendly, another green aspect of cabling and cable management is to improve air flow to boost cooling efficiency. Unorganized under- floor cabling results in air flow restrictions or blockages requiring HVAC and CRAC systems to work harder, consuming more energy to support cooling activities. Cabling should not block the air flow for perforated tiles on cool aisles or block movement of hot air upward in overhead conveyance systems. Reducing cable congestion has a positive PCFE impact by improving cooling efficiency as well as simplifying management and maintenance tasks.

Newer environmentally safe cabling that is physically smaller per diameter enables more cabling to be installed per footprint to help improve

PCFE issues. IOV and virtual connect technologies in blade centers for blade servers as well as high-density fan-in, fan-out cabling systems can further reduce the cabling footprint without negatively impacting networking or I/O performance for servers, storage, and networking devices. For cost-sensitive applications, shorter-distance copper cabling continues to be used, including for 10GbE, while fiber optic cabling continues to increase in adoption locally and on a wide area basis.

9.7   Summary

Networks can have a positive environmental impact by increasing telecommuting and reducing the number of vehicles on the road. Telecommunicating also requires that telecommunications networking and IT servers and associated networks be available and efficient. Moving forward, premium or low-latency lossless converged enhanced Ethernet (CEE) will complement traditional or volume Ethernet-based solutions leveraging various degrees of commonality. For storage-related applications that are not planned to be migrated to NAS or iSCSI, FCoE addresses and removes traditional issues about Ethernet-based TCP/IP overhead, latency, and nondeterministic behavior while preserving experience and knowledge associated with Fibre Channel and FICON tools and technologies.

To keep pace with improvements and new functionality being added to storage and servers and to boost efficiency, networks will need to do more than provide more bandwidth at a lower cost. There will be a need for faster processors, more interoperability and functionality, as well as technology maturity. More intelligence will be moving into the network and chips, such as deep frame and packet inspection accelerations, to support network and I/O QoS, traffic shaping and routing, and security, including encryption, compression, and de-duplication, among other factors. I/O and networking technology vendors include Adaptec, Adva, AMCC, Brocade, Ciena, Cisco, Dell, Emulex, Extreme, F5, HP, IBM, Intel, Qlogic, Riverbed, and Xsigo among others.

Action and takeaway points from this chapter include the following:

- Minimize the impact of I/O to applications, servers, storage, and networks.

- Do more with less, including improving utilization and performance.

- Consider latency, effective bandwidth, and availability in addition to cost.

- Apply the appropriate type and tier of I/O and networking to the task at hand.

- I/O operations and connectivity are being virtualized to simplify management.

- Convergence of networking transports and protocols continues to evolve.

There are several things that can be done to minimize the impacts of I/O for local and remote networking as well as to simplify connectivity. The best type of I/O is one that is efficient and as transparent as possible.

Chapter 10

Putting Together a Green and Virtual Data Center

It's not what you know; it's how you use it.

In this chapter you will learn:

- That not all data life-cycle and data access scenarios are the same
- How to compare and measure IT resources for different usage modes
- The importance of using capacity planning and data management
- How energy efficiency can be used to save costs or sustain growth
- Emerging roles and usage scenarios for virtualization beyond consolidation

The benefit of this chapter is to adjust company thinking about applying techniques and technologies including the many faces of virtualization to enable an economically efficient, productive, and environmentally friendly IT data center. Having looked at various issues, drivers, and components, let's shift gears and consider how to evolve toward a green and virtual data center that is able to sustain business growth.

10.1   Implementing a Green and Virtual Data Center

Building on the theme of closing the green gap between messaging and marketing rhetoric and IT issues, this chapter pulls together the various themes, topics, techniques, and technologies addressed in this book to enable a green and virtual data center. Although reducing carbon footprint and emissions are important and timely topics, the reality is that for many environments, public relations and environmental stewardship are agenda topics, but economic and business sustainment are the fundamental objectives.

Businesses of all sizes can and should invest holistically to reduce their environmental footprint. This goes beyond carbon emissions to include general recycling and improved resource usage efficiency. Improving on resource usage efficiency, including energy for powering and cooling IT resources, floor space, and facilities, contributes to the business's top and bottom lines and has the side benefit of helping the environment. For example, suppose that a manufacturing company can remain competitive and fit into an existing electrical power footprint and budget while reducing the number of BTUs needed from 20,000 to 15,000 per ton of product. This might be accomplished by either reducing energy consumption and associated costs or by increasing production to process more goods with the same energy footprint. Both of these methods will benefit the company both financially and environmentally.

The preceding example translates to similar scenarios for information factories, known as IT data centers:

- Energy consumption can be decreased to reduce cost and subsequent emissions.
- More work can be done or information stored using the same amount of energy.
- More work can be done or information stored using less energy reducing costs and emissions while boosting productivity and enabling future growth.

An important idea is to manage more data and more IT resources in the same, or even smaller, footprint, resulting in less energy usage and less floor space needed, and lower management costs. As shown in Figure 10.1, as the numbers of servers, storage, and networks—and the associated increase in data footprint, energy consumption, and costs—continue to rise, so will total infrastructure resource management (IRM) and hardware costs.

However, as shown in Figure 10.2, by improving efficiency and storing data within denser footprints without compromising performance or availability, and by improving how data, application, and IT resources are managed, IRM and energy costs can be used more efficiently, enabling more

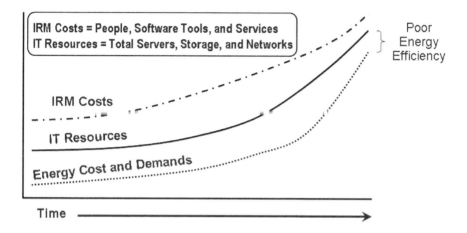

Figure 10.1   IT Resource Inefficiency Means Rising Costs and Complexity

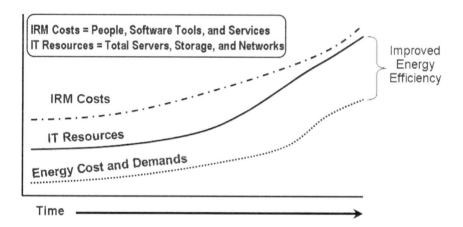

Figure 10.2   Shifting to an Efficient and Productive Information Model

work to be done and more data to be stored. This, in turn, will help the environment while sustaining business and economic growth. The objective is to decrease the energy efficiency gap and to manage more with less.

By increasing the efficiency of how data and applications, along with the associated IT resources (servers, storage, networking, and facilities), are managed, more work and data can exist in a smaller footprint.

10.2   PCFE and Green Areas of Opportunity

Just as there are many different issues that fall under the green umbrella, there many different approaches and opportunities to address them. These include PCFE, cost reduction, and sustainment of growth. Table 10.1 summarizes various action items and opportunities that were shown in Chapter 1, in Figure 1.5.

Table 10.1   Action Items and Opportunities to Address PCFE and Green Issues

Activity	Description	Opportunity
Financial incentives	Rebates, low-cost loans, grants, energy affiance certificates	Offset energy expenses and technology upgrade costs with rebates or leave potential money on the table
Metrics and measurements	Total energy usage and resulting footprint, along with how much work is being done or data is being stored per unit of energy consumed	Provide insight and enable comparison of productivity and energy efficiency to gauge improvement, success, and environmental impact
Infrastructure resource management	Leverage best practices, protecting and securing data and applications while maximizing productivity and resource usage using various technologies	Reduce IRM complexity and costs, boost productivity by doing more with less while enhancing IT service delivery, including performance and availability
Mask or move problems	Outsource; use managed service providers, software as a service (SaaS), or cloud services; buy carbon offsets to meet emission tax scheme (ETS) requirements as needed	Utilize carbon offset credits to comply with ETS when needed; leverage lower-cost services when applicable without compromising IT service delivery
Consolidation	Leverage virtualization in the form of aggregation of servers, storage, and networks; consolidate facilities, applications, workload, and data	Reduce physical footprint to support applications and data conducive to consolidation, balancing savings with quality of IT service delivery
Tiered resources	Various types of servers, storage, and networking components, sized and optimized to specific application service requirements	Align technology to specific tasks for optimum productivity and energy efficiency; balance performance, availability, capacity, and energy

Table 10.1   Action Items and Opportunities to Address PCFE and Green Issues

Activity	Description	Opportunity
Reduce data footprint	Archive, compression, de-duplication, space-saving snapshots, thin provisioning, and data deletion	Eliminate un-needed data, move dormant data offline, compress active data, maximize density
Energy avoidance	Power down resources when not in-use using various energy-saving modes	For applications and IT resources that are conducive to being powered down, turn off when not in use, including workstations and monitors
Boost energy efficiency	Upgrade to newer, faster, denser, more energy-efficient technologies; leverage tiered servers, storage, and networks	Maximize productivity and amount of work done or data stored per watt of energy in a given footprint, configuration, and cost
Facilities tune-up	Leverage precision cooling, review energy usage, and assess thermal and CRAC performance. Eliminate halon and other environmentally unfriendly items.	Reduce the amount of energy needed to cool IT equipment, doing the same or more work while reducing energy costs and usage
Environmental health and safety	Recycle, reuse, reduce, eliminate e-waste and hazardous substances	Comply with current and emerging regulations, including those involving removal of hazardous substances

10.2.1   Obtain and Leverage Incentives and Rebates

If you have not already done so, take advantage of available energy-efficiency rebates or incentive programs. Talk with energy providers to learn what existing, emerging, or custom programs your organization can utilize to reduce energy bills, offset energy spending, or help to fund future upgrades and technology refresh. Not taking advantage of every available program or rebate is the same as walking away from money on the table.

10.2.2   Best Practices and IRM

Virtualization can be used in many ways, including consolidation, abstraction, and emulation, to support load-balancing, routine maintenance, and business continuity/disaster recovery (BC/DR. In Figure 10.3, a traditional BC/DR environment is shown on the left, with dedicated physical resources

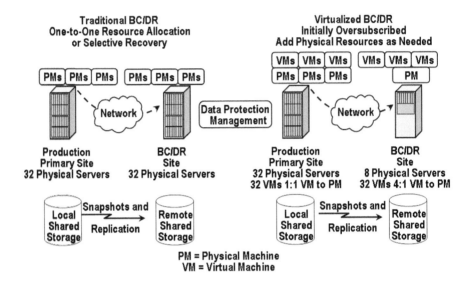

Figure 10.3   Leveraging Virtualization to Enable High Availability and BC/DR

or selective applications being recovered. Challenges include dedicating extra hardware on a one-to-one basis and deciding which servers and applications should be recovered to available physical resources.

Maintaining BC/DR plans is complex, including testing of configuration changes as well as associated costs of hardware, software, and ongoing operational costs of power, cooling, and floor space. Other issues and challenges include difficulties in testing or simulating recovery for training and audit purposes as well as inefficient use of available network bandwidth, inhibiting the amount of data that can be moved in a timely fashion.

The right side of Figure 10.3 shows a solution that leverages virtualization for abstraction and management in which each physical server is converted to a virtual machine (VM). However, the VM is allocated a physical machine (PM) such as a server or server blade on a one-to-one basis. In case of a disaster, or for BC or training and testing, multiple VMs can be recovered and restarted on a limited number of PMs, with additional PMs being added as needed to boost or enhance performance. Reduce data footprints to boost data movement and data protection effectiveness using a combination of archiving, compression, de-duplication, space-saving snapshots, data replication, and bandwidth optimization. Data protection management tools can manage snapshots, replication, backup, and associated functions across serves, storage, network, and software resources.

Benefits of server virtualization include more efficient use of physical resources; the ability to dynamically shift workloads or VMs to alternate hardware for routine maintenance, high-availability, or BC DR purposes; support for planned and unplanned outages; and the enablement of training and testing of procedures and configurations. In addition, the approach shown on the right side of Figure 10.3 can also be used proactively for routine IRM functions—for example, shifting applications and their VMs to different physical severs either on-site or off-site during hardware upgrades or replacement of serves or storage.

A variation of Figure 10.3 uses virtualization for abstraction to facilitate provisioning, configuration, and testing of new server and application deployments. For example on the left side of Figure 10.4, multiple VMs are created on a physical machine, each with a guest operating system and some portion of an application. During development and testing and to support predeployment IRM maintenance functions, the various applications are checked on what appears to be a separate server but in reality is a VM.

For deployment, the various applications and their operating system configurations are deployed to physical servers, as shown in the right side of Figure 10.4. There are two options: Deploy the applications and their operating system on VMs allocated one to one with physical servers; or convert the VMs, along with guest operating systems and applications, to run on physical servers without a virtualization layer. With the first

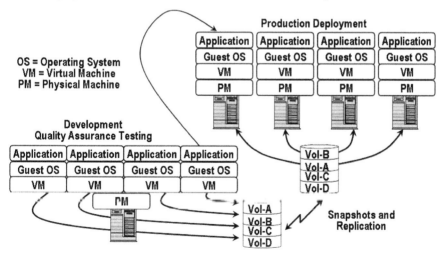

Figure 10.4   Utilizing Virtualization for Abstraction and Server Provisioning

option, virtualization is used for abstraction and management purposes as opposed to consolidation. An underlying VM enables maintenance to be performed as well as providing the ability to tie into a virtualized BC/DR scheme as shown in Figure 10.4.

10.2.2.1 Assessing IT Resource Usage Effectiveness and Capacity Planning

To avoid future capacity shortages or service outages due to lack of sufficient resources and to ensure existing resources are being used optimally, create a storage capacity plan. A storage capacity plan can be as simple as outlining roughly when and how much storage performance, availability, and capacity will be needed. Capacity planning is the practice of knowing what resources are available, how resources are utilized, and what resources will be needed in the future. For IT environments of all sizes, capacity planning involves knowing what IT resources or servers, storage, and networking resources are present, how they are being used to deliver service at a given level of quality, and when future upgrades of hardware, software, facilities, and related services will be needed.

IT capacity plans for forecasts can be very simple or extremely detailed, using models and sophisticated projections; in either case, however, they need to align with growth and business plans. In the simplest format, a capacity plan considers what resources currently exist, such as the number of servers, amount of memory, number and type of networking ports, software license, storage performance capabilities, and space capacity, as well as how they are being used. More specifically, a simple capacity plan indicates what percentage of the servers are busy at what times and for how long, how much memory is being used, how many megabytes or gigabytes of data are being moved per second or hour or day, as well as how much storage capacity space is being used. Additional information might include response time or some other measure of productivity along with availability.

Capacity planning should start simply, showing results and benefits, and evolve to more complex plans encompassing various approaches. Tools such as modeling, reporting, and forecasting can be added with an eye toward utilization as well as performance, response time, and availability. Plans should cross different IT IRM domains—that is, servers, networks, storage, and input/output (I/O) as well as facilities PCFE.

Establish a baseline for your environment and applications, including what is normal or typical performance. For example, if you know what typical I/O and throughput rates are for various devices, what the common error rates are, average queue depths and response times, you can use this information to make quick comparisons when things change or if you need to look into performance issues.

Capacity planning needs to consider planned and seasonal workload increases—holiday and back to school shopping periods, spring break and summer vacation travel—and other events that may cause a surge in activity. For larger environments, automated and intelligent tools can be used to track deviations from normal performance as well as to help pinpoint problem areas quickly. Part of establishing a baseline is gaining insight at different levels of what servers and devices are the "top talkers" in generating traffic, with various levels of detail including specific LUNs or volumes, ports, protocols, and even applications on servers.

Unless an inventory or list of resources is available, a good starting point for capacity planning is to take stock and assess what resources currently exist and how they are being used—for example, how much storage exists, how it is configured, how much of it is usable and allocated, who is using the storage, and when the storage is being used. It is also important to look at storage performance, including I/O rate, types of I/Os (reads or writes, random or sequential) if available, data transfer rates (how much data is being moved), and availability. In identifying or assessing storage and server resources, a simple guideline to keep in mind is performance, availability, capacity, and energy consumption (PACE).

In assessing performance, keep response time and latency in perspective. A by product of assessing available storage resources may be the discovery of unused or unallocated storage as well as data that may be eligible for archiving. The result can be that additional storage capacity can be recovered for near-term growth.

Items to consider when performing an assessment include the locations, numbers, makes, and models of servers, storage, and networking devices, along with their configurations and what applications they support. When assessing configurations, include what software and versions are installed, numbers and types of I/O and networking interfaces, protocols in use, amount of memory or cache, and quantity of processors or controllers. For servers, include the speed and amount of memory along with the number of processor cores. For storage, include the quantities, types, and capacities of disk drives as well as how they are configured, including RAID levels, size and number of LUNs, and how the storage is allocated to different servers or applications. For networking devices, include the numbers and types of ports and specialized routing or application blades.

An assessment, in addition to logical and physical configuration, should include how the resources are being used—for example, CPU and memory utilization of servers over different time intervals along with application workload indicators such as transactions, file requests, email messages, or videos served in a given time period. In addition to CPU and memory usage for servers, look at queue depths and response times as well as availability to gauge the effectiveness of service delivery combined with application- or business-focused service delivery metrics. For storage systems, in addition to looking at raw and usable or effective storage capacity, look at performance- or activity-related metrics for active storage, including IOPS, bandwidth, video streams, file requests, as well as response time and availability. For networking devices, look at the utilization of ports along with errors and congestion indicators.

Energy usage and cooling demands for IT resources should also be captured, using either nameplate, vendor-supplied, or actual measurable information. Where possible, link the energy usage and cooling requirements to actual work being done or usage scenarios, either automatically via IRM tools or manually, to maintain a profile of efficiency and effectiveness. For example, armed with the applicable information, you should be able to piece together how much energy is needed to power and cool all the IT resources needed to perform backups, support BC/DR, and enable routine maintenance, as well as to enable basic IT service delivery of online and other applications and data.

Other items to include when assessing or establishing a configuration baseline and profile are application- or system-specific RTO and RPO requirements, including how and when data is protected and to where. Service requirements and data-retention policies combine to determine how long and where data should be kept to facilitate archiving of data off of primary or active storage to near-line or offline storage if you cannot actually delete data.

Armed with an inventory of available resources and how they are being used, another component is to determine how resources have been used in the past. Part of determining the future is to look at previous trends and issues and at how well previous forecasts have aligned with actual usage. In addition to looking into the past, look into the future, including assessing business and application growth rates. Aligning future growth plans, current activity, and past trends can help you formulate future upgrade and capacity needs.

Use the information from the assessment, considering inventory, growth plans, and previous usage information, to put together a forecast. The forecast can be as simple as taking current usage (space and performance), assuming that current response-time objectives are being met, and applying growth factors to them. Of course, a capacity plan and forecast can be much more sophisticated, relying on detailed models and analysis maintained by in-house or external sources. The level of detail will vary depending on your environment and specific needs. The tricky part, which has resulted in capacity planning being called "part science and part black magic," is in knowing how to determine what growth rates to use as well as applying "uplift factors." An uplift factor is a value to account for peak activity, unanticipated growth, and "swag" or buffer space.

10.2.2.2   Securing IT Resources and Data

Data security in general, including logical (authorization, authentication, encryption, passwords) and physical (locked doors, surveillance, or access control), has traditionally been associated with large-enterprise applications or with environments with sensitive data. The reality is that, given the increase in reliance on information and awareness of concerns about data privacy, data security is an issue of concern for all environments.

Not all data is physically transported when it is moved between locations, as there is a growing trend toward moving data electronically over networks. Data is moved over networks using backup, archiving, real-time, or time delayed replication and file-copy techniques. If you are looking to provide network-based data movement capabilities to your customers, you will want to take into consideration available and accessible network bandwidth, latency, and fault isolation or protection, not to mention cost and data encryption. Part of the network cost for moving data is the up-front cost of acquiring the network service as well as special equipment and recurring monthly network costs.

Physical security items include:

- Physical card and ID for access, if not biometric access card for secure facilities
- Secure and safe disposition of storage media and assets

- Secure digital shredding of deleted data, with appropriate audit controls
- Locked doors to equipment rooms and secure cabinets and network ports
- Asset tracking, including portable devices and personal or visiting devices
- Limits or restrictions on photo or camera usage in and around data centers
- Low-key facilities without large signs advertising that a data center is here
- Protected (hardened) facility against fire, flood, tornado, and other events
- Security cameras or guards

In addition to addressing physical security of data, application, and IT resources, logical security includes:

- Encryption key management
- Forced regular changes of passwords combined with digital rights management
- Authentication of user credentials and authorization of individual rights
- Logical storage partitions and logical or virtual storage systems
- Tamper-proof audit trails and logs of who accessed what, when, and from where
- Encryption of data at rest (on disk or tape) or in flight (transmitted over network)
- Secure servers, file-systems, storage, network devices, and management tools

10.2.3   Implement Metrics, Measurements

Moving forward, in addition to looking at performance and availability, power, cooling, and energy consumption will also become factors, both from an energy-efficiency perspective (energy conservation) and from an effectiveness perspective (how much work is and can be done per watt of energy).

For example, look at IOPS or bandwidth per watt or kilowatt-hour to gauge activity as well as capacity for a given footprint, configuration, and price point. In addition to measuring IOPS, bandwidth, and capacity utilization, other metrics include response time or latency and availability. Application or workload activity metrics include transactions, files or emails, videos, photos, or other useful work done per time interval and energy consumed.

Although there are many different types of metrics and measures addressing performance, availability, capacity, and energy in various combinations, these need to be looked at in two key ways. The first way to look at data center resources is from an active standpoint; the second is from an idle, inactive, or lower power, energy-saving view. Both active and idle metrics have their place in considering resources for different usages or points in time. For example, storage for offline data should be measured on an idle or inactive basis, because that is where most of its time and usage is spent. On the other hand, a server, storage, or networking device that is always on and in active use should be looked at from the viewpoint of how much work can be done and the energy used while it is active.

In between these two extremes are mixed usage cases, where servers, storage, and networking devices are busy part of the time and idle, inactive, or in low-performance modes during other times of the day, week, or month. During busy times, the resources should be measured on how efficient they are at doing useful work per energy used. During inactive time, look at how energy-efficient the devices are when they are idle or in low-power mode. Most servers, networking, and storage devices (except those for offline archiving) have as a primary objective and business justification doing some amount of work. Thus, while it is good to know how efficient the devices are in low-power mode, it is equally, if not more, important to know how energy-efficient the devices when actually doing work .

10.2.4 Mask-or-Move Issues

There are several approaches to masking or moving IT PCFE and green issues. For example, to comply with new ETS requirements to make up for excessive energy use and subsequent carbon footprint, carbon offsets or carbon taxes can be bought or paid. This may be an applicable business decision on a near-term basis while efforts are underway to reduce emissions or energy consumption, not only for IT data centers, but also for organizations or businesses as a whole. There are also many carbon trading schemes, similar to other commodity trading systems.

Carbon offsets, for the most part, outside of areas and industries that fall under an ETS, are voluntary in nature or make for good public relations. An alternative to paying for a carbon offset is to improve on energy efficiencies and reduce emissions, which may also qualify for financial incentives or rebates. When put into business terms, what is more appealing—paying for offset credits or potentially getting money back, or a discount on energy usage, while achieving energy efficiency and improved productivity in the process? Of course, case-by-case business models and situations need to be analyzed to see if spending money actually saves money on a near-term or long-term basis.

Approaches to mask or move PCFE issues include:

- Buy carbon offset credits
    - Enables compliance for energy trading schemes
    - Buys time until efficiency (ecological and economic) are achieved
    - Adds cost to business when not part of regulatory compliance
    - Does not address energy efficiency but is, rather, a deferral
- Leverage managed service, SaaS, cloud, and hosting services providers
    - Move problems and issues elsewhere
    - Take advantage of lower-cost services or available PCFE resources
    - Supplement existing environments using lower-cost services
    - Fully or partially outsource some applications and services
- Build new facilities, expand or remodel existing facilities
    - Expensive and time-consuming; strategic versus tactical
    - Part of solution when combined with other techniques
    - Look for sites with abundant power generation and transmission capabilities

10.2.5   Consolidation

Consolidation can be used to improve the energy efficiency of underutilized resources, reducing PCFE footprints across servers, storage, networks, and facilities. Excess capacities from servers, storage, and networks as a result of consolidation can be used to support growth or enhance existing applications. Surplus technologies can also be removed, making way for newer, faster, denser, and more energy-efficient technologies to further reduce costs

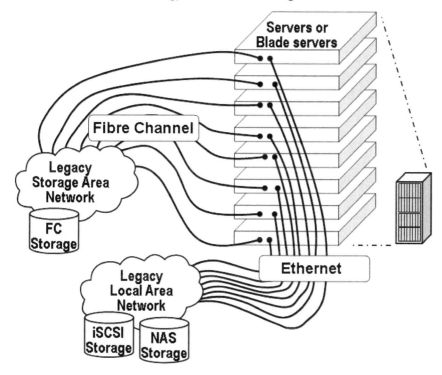

Figure 10.5   Proliferation of I/O and Networking Adapters and Cabling

and PCFE impacts while sustaining growth.

Figure 10.5 shows an example of a blade server center or high-density rack servers along with I/O and networking connectivity for storage to Fibre Channel SAN and general networking along with iSCSI and NAS storage via Ethernet. For high-availability environments, two HBAs and two NICs would be used for redundancy. In this example, eight servers each have at least two adapters or NICs, one for Fibre Channel and one for Ethernet.

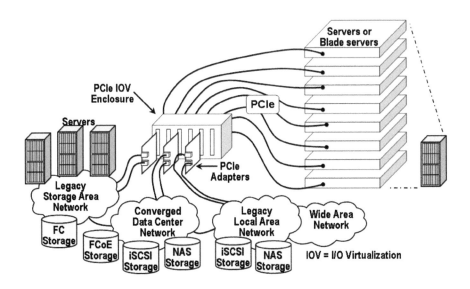

Figure 10.6 Consolidated I/O and Networking Connectivity

Building on the example in Figure 10.5, and assuming a non-high-availability configuration, Figure 10.6 shows a simplified I/O and networking configuration using PCIe IOV along with converged networks. In Figure 10.6, instead of 16 PCIe adapters for a non-high-availability configuration as in Figure 10.5, three adapters are shown.

In Figure 10.5, instead of dedicated PCIe adapters in each server, using PCIe IOV, a PCIe extension connection extends from each server to a PCIe adapter enclosure using multiroot IOV mode adapters that can be shared across physical servers, reducing the number of I/O and networking components. With increased scaling of servers, the connectivity components including adapters, optical transceivers, cabling, and switches can be reduced using I/O and networking consolidation.

10.2.6 Reduced Data Footprint

An approach to reducing data footprint is to prevent or avoid a growing data and storage presence by establishing quotas and curbs on growth as well as aggressive data deletion polices. However, if these draconian approaches are not applicable, then several approaches can be used in various combinations to manage an expanding data footprint more effectively. Where possible, identify data that is truly no longer needed and, if possible, archive to some other long-term media such as magnetic tape, optical drives, or removable magnetic disk drives before deleting it.

For inactive data that will or might be needed in the future or that falls under regulatory compliance or other retention needs, make a copy or multiple copies and either move them offline to removable media or leverage a managed service, SaaS, or cloud service provider to park the data until it may be needed. Archiving can be applied to both structured databases, including actual data as well as associated business rules and relationships, and to email and other unstructured data. Data that is archived should be compressed into a denser format for storage and, if necessary, maintained in a preserved state to comply with regulatory or legal requirements.

Compression should be used for both online active data and for backup, data protection for BC/DR, and archiving. In the big picture, if, for example, only about 5–10% of data is changing and thus being backed up, that leaves 90–95% of the remaining active online data as an opportunity for data footprint reduction. Of course, the amount of changing data will vary with the specific applications and environments. The key point here is that, in addition to focusing on reducing data footprints associated with backup, look across all tiers and categories of data and storage usage to achieve even greater benefits from a reduced data footprint.

Compression can be implemented on servers in applications; for example, using database or email-based real-time compression or via operating system and third-party software. Real-time compression can also be used in the I/O and network data path, using appliances that can compress and decompress data on the fly without noticeable performance impacts. For example, while intuitively there should some amount of overhead to compress or decompress on the fly, the overhead should be offset by the reduced footprint of data that needs to be written to or read from a storage system and transferred over a network. Compression is also often implemented in storage devices such as magnetic tape drives to reduce the data footprint, improve the density of data being written to the medium, and boost effective performance to the media.

A smaller data footprint also means that more data can fit into existing storage system cache, boosting performance and maximizing cache effectiveness. With a smaller on-disk data footprint, for subsequent data access, more data can be prefetched as part of the read-ahead algorithms that are common on many storage systems. Consequently, frequently accessed data is more likely to be readily available in a storage system protected cache, boosting application performance and user productivity. The result is maximum usage of existing storage systems capabilities in an energy-efficient manner and doing more with less, including boosting the available effective cache for performance on storage systems.

For read operations, less data needs to be retrieved, making read-ahead caches and algorithms more effective by transferring more data in less time, enabling more usable performance on a storage system. The end result is an improved effective storage system with better performance on reads and writes, with a smaller data footprint requiring less storage. The benefits are that more performance can be obtained from existing storage systems while boosting storage system cache effectiveness and storing more data in a smaller footprint. For example, instead of using 750-GB, 7.2K-RPM SATA disk drives, a NAS filer could be configured with 300-GB, 15.5K FC or SAS disk drives that have a similar power draw on a spindle-to-spindle basis.

Compression ratios and wire speed performance rates can be interesting; however, they may not be relevant. Look at the effective compression rates and effective compression impact of a solution. For example, at a 20-to-1 ratio on 20 TB of backup data, reducing it to 1 TB is impressive (e.g., elimination of 19 1-TB disk drives). Assuming a 2- or 3-to-1 or better compression ratio, the NAS filer would have the similar or better capacity with the same number of hard disk drives. However, the performance benefit of the disk drives that are twice the performance in RPMs would yield better IOPs as opposed to the slower SATA drives that could result in a performance bottleneck. What is even more significant is that with as little as a 3-to-1 compression ratio, for 1,000 TB of online primary data, the data footprint could be reduced to about 333 TB. Real-time compression that provides an effective performance benefit also enables the equivalent savings of 666 1-TB disk drives. The impact is much larger if smaller, high-capacity disk drives are being used.

The net result is the ability to delay installing more storage or to increase the effective capacity of existing and new storage while enhancing performance. This approach, even on a smaller scale, can be used to maximize available power and floor space during technology upgrades. For example, by compressing active data, data from other storage systems can be consolidated to free up floor space and power during instillation and migration to newer technologies.

Data de-duplication to remove or eliminate recurring data and files is an emerging technology that is getting a lot of attention and is being adopted for use with backup and other static data in entry-level and mid-range environments. De-duplication works well on data patterns that are seen as recurring over time—that is, the longer a de-duplication solution can look at data streams to compare what it has seen before with what it is currently seeing, the more effective the de-duplication ratio can be. Consequently, it should be little surprise that backup in small to medium-size environments is an initial market "sweet spot" for de-duplication, given the high degree of recurring data in backups.

The challenge with de-duplication is the heavy thinking needed to look at incoming data or data that is being ingested and determine if it has been seen before, which requires more time, thinking, or processing activity than traditional real-time compression techniques. While the compression ratios can be larger with de-duplication across recurring data than with traditional compression techniques, the heavy thinking and associated latency or performance impacts can be larger as well. De-duplication has great promise and potential for larger-scale environments. that are currently utilizing magnetic tape. Over time, as de-duplication algorithms improve and processing power increases, the heavy thinking associated with de-duplication should be addressed to reduce the time and latency required while supporting much larger scaling with stability data footprint reduction.

A hybrid approach for de-duplication is policy-based, combining the best of both worlds and providing IT organizations the ability to tune de-duplication to given quality of service (QoS) needs. Policy-based data de-duplication solutions provide the flexibility to operate in different modes, including reducing duplicate data during ingestion, on a deferred or scheduled basis or with de-duplication turned off and on as needed. The benefit of policy-based de-duplication is flexibility to align performance, availability, and capacity to meet different QoS requirements with a single solution. For example, for performance- and time-sensitive backup jobs that must complete data movement within a specific time frame, policy-based de-duplication can be enabled to reduce backup windows and avoid performance penalties. For usage scenarios where the emphasis is on utilizing as little target, backup, or archive storage space as possible without concern for time or performance, than immediate de-duplication can be chosen. Yet for other applications that gain no immediate benefit from data de-duplication, such as large database dumps or backups, de-duplication can be disabled.

Going forward, de-dupe discussions or "de-bates" will shift away from the pros and cons of various algorithms and who has the best de-duplication ratio to who has the most flexibility and scalability. This flexibility will mean the ability to align data footprint reduction to different performance and capacity needs as well as the ability to coexist with tape for destaging data to offline media with replication. Scaling will include the ability to support larger raw capacity, levering robust and reliable storage systems while enabling performance to support larger environments—including those still dominated by disk-to-tape and disk-to-disk-to tape solutions.

10.2.7   Tiered Servers, Storage, and I/O Network Access

A current-generation SAS or Fibre Channel 15.5K or higher disk drive with one-third or one-fourth the capacity of a high-capacity disk drive may consume the same or even less power on a drive-per-drive basis. The caveats are to compare like drive types and form factors and to remember that there will be some variances due to power management, among other items. However, generally speaking, as an example, a current-generation 146-GB, 15.5K, 4GFC, 3.5-inch Fibre Channel disk draws less power than a previous-generation 146-GB, 15K, 3.5-inch Fibre Channel disk because of technology and manufacturing improvements. Similarly, a current-generation 146-GB, 15.5K, 4GFC, 3.5-inch disk drive draws about the same or possibly less, depending on specific make and model, than some 5,400-RPM or 7,200-RPM, 750-GB or even larger disk drives.

While this may seem hard to believe given the industry talk about large-capacity SATA disk drives being more energy-efficient, the reality is that on a capacity basis, large-capacity disk drives do provide more capacity per watt of energy in a given footprint and for a given cost point. However, a high-performance disk drive provides more performance in terms of IOPS or bandwidth per watt of energy for a given capacity and price point in the same footprint.

For high-performance storage needs, use energy-efficient current-generation fast disk drives to enable more work to be done instead of using multiple slower disk drives. Applications that need large amounts of storage capacity with low performance needs should use high-capacity SAS and SATA disk drives. Meanwhile, extreme high performance and I/O-intensive applications can benefit from using RAM or RAM+FLASH SSDs for relatively small amounts of data needing large amounts of performance and low latency.

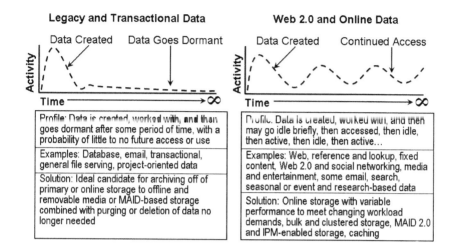

Figure 10.7   Changing Access and Data Life-Cycle Patterns

Many strategies or marketing stories are built around the premise that shortly after data is created, the data is seldom, if ever, accessed again. This model lends itself to what has become known as *information life-cycle management* (ILM), where data can and should be archived or moved to lower-cost, lower-performing, and high-density storage or even deleted where possible. Figure 10.7 shows an example, on the left side of the diagram, of the traditional data lifecycle, with data being created and then going dormant. The amount of dormant data will vary by the type and size of the organization as well as the application mix; for example, from 55% to 80% of data may be dormant. However Figure 10.7 also shows a different data life cycle, one more common for online and Web-based data, including Web 2.0 social networking and collaborative applications. Unlike traditional data life-cycle models that favor offline and hybrid data movement and archiving solutions, other data life-cycle models like that shown in Figure 10.7 may have a very low percent—perhaps single-digit amount—of data that is conducive to traditional ILM models. Instead, this category of data needs to remain online on low-power, moderate-performing, high-capacity, and low-cost storage that is predominantly read-accessed.

Unlike traditional data life cycles, in which data goes dormant after a period of time, on the right side of figure 10.7, data is created and then accessed on an intermittent basis with variable frequency. The frequency between periods of inactivity could be hours, days, weeks, or months and, in some cases, there may be sustained periods of activity. A common example is a video or some other content that is created and posted to a website or social networking site such as Facebook, Myspace, or YouTube. Once the content is discussed, it may not change, and additional comment and collaborative data can be wrapped around the data as additional viewers discover and comment on the content. Unlike traditional data life-cycle models, in which data can be removed after a period of time, Web 2.0 and related data needs to remain online and readily accessible. Solution approaches for the new category and data life-cycle model include low-cost, relatively good-performing high-capacity storage such as clustered bulk storage.

Given that a large percentage of data in the Web 2.0 and related data life-cycle model is unstructured, NAS-based storage solutions including cloud and managed service offerings with file-based access are gaining in popularity to address these needs. To reduce cost, a growing trend is to utilize clustered NAS file systems that support NFS and CIFS for concurrent large and small I/Os as well as optionally pNFS for large parallel access of files.

In addition to using software-based clustered file systems deployed using general-purpose servers with varying amounts of memory, processors and I/O adapter capacities including blade center servers are being implemented. Another trend is to utilize small-form-factor energy-efficient SAS disk drives for performance and small-form-factor, high-capacity SATA disks in the same external RAID enclosures. Some low-cost solutions combine PCIe SAS-based RAID adapter cards installed into servers running clustered NAS file system software and leverage high-density, low-cost, high-capacity JBOD disk enclosures.

10.2.8   Energy Avoidance—Tactical

There is a lot of focus on energy avoidance, as it is relatively easy to understand and it is also easy to implement. Turning off the lights, turning off devices when they are not in use, enabling low-power, energy-savings or Energy Star® modes are all means to saving or reducing energy consumption, emissions, and energy bills. Ideal candidates for powering down when not in use or inactive include desktop workstations, PCs, laptops, and associated video monitors and printers. Turning lights off or implementing motion detectors to turn lights off automatically, along with powering off or enabling energy-saving modes on general-purpose and consumer products has a significant benefit.

Given the shared nature of their use along with various intersystem dependencies, not all data center resources can be powered off completely. Some forms of storage devices can be powered off when they are not in use, such as offline storage media for backups and archiving. Technologies such as magnetic tape or removable hard disk drives that do not need power when they are not in use can be used for storing inactive and dormant data.

For data center devices that cannot be powered down completely, consolidation and upgrades to newer, more energy-efficient technologies that are 80%-plus efficient are an option, as are intelligent power management (IPM), adaptive power management (APM), adaptive voltage scaling (AVS), second-generation MAID or MAID 2.0, and other variable-power modes that enable some work to be done while consuming less power during slow or inactive times.

10.2.9   Energy Efficiency—Strategic

Avoiding energy use is part of an overall approach to boosting efficiency and addressing PCFE challenges, particularly for servers, storage, and networks that do not need to be used or accessible at all times. However, for applications and data that need to be available and accessible, boosting energy efficiency is an important and strategic topic. Simply put, when work needs to be done or information needs to be stored or retrieved or data moved, it should be done so in the most energy-efficient manner aligned to a given level of service.

Instead of measuring how much power is avoided by turning equipment off, for active applications, data, servers, storage, and networks, the metrics become how much data can be stored in a given footprint yet still be accessible in a given time frame to meet service-level objectives. Another measurement is how much work, how many transactions, how many files or email requests, how many users or video streams can be processed in a given amount of time, as fast as possible with the least amount of energy being used to power and cool equipment.

General approaches to boost energy efficiency include:

- Do more work while using the same or less power.
- Leverage faster processors/controllers that use the same or less power.

- Consolidate slower storage or servers to a faster, more energy-efficient solution.

- Use faster disk drives with capacity boost and that draw less power.

- Upgrade to newer, faster, denser, more energy-efficient technologies.

- Look beyond capacity utilization; keep response time and availability in mind.

- Leverage IPM, AVS, and other modes to vary performance and energy usage.

- Manage data both locally and remote; gain control and insight before moving problems.

- Reduce data footprint impact, enabling higher densities of stored data.

10.2.10   Facilities Review and Enhancements

Conduct an energy and thermal assessment of data center facilities to validate energy usage along with opportunities for improvement near-term and longer-term. Look into how energy is being use for cooling, and consider opportunities to improve the cooling and heat-removal efficiencies of HVAC and CRAC resources. Investigate what server, storage, and networking resources can be upgraded to newer, faster, and more scalable and energy-efficient technologies. As part of a facilities review and tune-up, look into replacing legacy halon fire-suppression systems with newer, more environmentally and health-safe technologies, as well as advanced smoke and fire detection capabilities.

10.2.11   Environmental Health and Safety; E-Waste; Recycle, Reuse, Reduce

Activities to address environmental health and safety (EHS) and related issues include recycling materials in office areas as well as recycling electronic equipment and shipping containers. Discuss with current or prospective vendors, particularly those who want to tell you how green they or their solutions are, what their green ecosystem and supply chain looks like. Have the vendors explain what they do internally with regard to EHS and material management systems as well as how they manage their suppliers. Look for vendor participation in broader initiatives beyond industry trade associations, such as the Environmental Protection Agency's Climate Leaders and Carbon Disclosure Project as an indicator of how serious they are about actually being green instead of merely being seen as being green.

Join voluntary programs to take action on addressing EHS items including EPA Climate Leaders and Energy Star for Data Center programs and Carbon Disclosure Project, among others. Deploy complaint technology for removal of hazardous substances, eliminating older, less energy-efficient, and less environmentally healthy or friendly technologies and materials. Reduce, reuse, and recycle IT equipment as well as office material. Explore alternative energy source opportunities in addition to using energy-efficient technologies.

10.3   Summary

It's not what you know; it's how you use it. With that in mind, keep in perspective all of the different techniques, technologies, and best practices discussed in this book that can be applied in various combinations to address PCFE or green issues. The principal takeaway for this chapter is to understand the different aspects of being green along with the technologies and techniques that can be used for different situations.

Action and takeaway points from this chapter include the following:

- Data and storage access and life-cycle models are changing.
- Metrics are important for gaining insight into resource efficiency.
- Virtualization can be used to enhance productivity and IT agility.
- Capacity planning should cover servers, storage, networks, and facilities.
- There is a difference between energy avoidance and energy efficiency.
- A green and virtual data center can reduce costs.
- A green and virtual data center can reduce energy usage and cmissions.
- A green and virtual data center can enable growth.
- A green and virtual data center can maximize or extend IT budgets.

Chapter 11

Wrap-up and Closing Comments

A little less conversation, a little more action please.—Elvis Presley

There are many facets to being green, and many technologies and solutions to address different issues and challenges. The concept of the "green gap" was introduced in Chapter 1, and various solutions and techniques were explored in subsequent chapters. This chapter provides an overview and some final comments.

11.1   Where We Have Been

I talk with IT data center professionals from around the world on a regular basis. When I ask whether they are pursuing green initiatives or are working under green mandates, most respond in the negative, though the number of positive answers is increasing. When I ask whether they are considering green initiatives, a few more respond positively, varying by region. The challenge is to close the gap between green messaging and green washing themes and make the link to solving IT data center issues including power, cooling, floor-space, and environmental health and safety (PCFE).

When I ask data center professionals whether they have current or near-term PCFE issues, the number of positive responses increases dramatically. The overall response varies by region and countries around the world; however, the responses signal that the green gap needs to be addressed to enable vendors to get their solutions to prospective customers, and customers to understand vendors' capabilities to address their current issues and needs.

There are many different messages about what it means to be green or to go green as well as why IT data centers need to become green. Green can mean energy avoidance or conservation as well as energy efficiency, reducing emissions and carbon footprint, reducing energy costs, and addressing environmental health and safety (EHS) initiatives, e-waste and general waste recycling, and removal of hazardous substances (RoHS), among other things. There is plenty of room for discussion and debate on the real or perceived environmental impact of IT data centers and their associated technology, including servers, storage, networks, and facilities, as well as how they are managed.

Many green-related discussions tend to center around how much of a carbon or emissions footprint IT data centers are actually responsible for on a regional, national, or global basis. Environmental sustainment and stewardship are important, but so, too, is supporting the increasing reliance on IT-delivered services in an economical manner. The central theme of creating a "green and virtual data center" should be to enable a cost-effective IT organization that is able to sustain business growth and application enhancement in a flexible and resilient manner in an economical and environmentally friendly manner. By addressing energy efficiencies of how IT resources are used moving forward to support growth, organizations can address other issues, including fitting into existing or future electrical power footprints, budget constraints, and environmental and emissions regulations, all while helping both the bottom line and the environment.

11.2   Where We Are Going—Emerging Technologies and Trends

With history as an indicator, it is safe to assume that, at least for the foreseeable future, demands and reliance on information systems and data availability will continue to grow. Likewise, it is safe to assume that data growth rates, in terms of both sheer volume and the diversity of digital data types, will also continue to expand. Consequently, ensuring adequate IT resources to support and sustain growth is essential. Technology continues to emerge and mature in general, and for IT specifically, to support next-generation virtual data centers. What was new and next-generation technology a few years ago is now being adopted in data centers to help enable productivity as well as address various issues. Some of today's new technologies and solutions will evolve and mature, living up to their hype in the not so distant future, while other technologies long ago declared dead will continue to have useful purposes.

What are some of the technologies that are emerging, evolving, and maturing to help transition today's IT environments and data centers to next-generation green and virtual data centers? Some are hardware, some are software and some network-based, some are combinations and it may not be clear if they are a service, software, network, I/O, server, or a storage solution crossing multiple technology domains and infrastructure resource management (IRM) discipline areas. Work is being done with advanced voltage scaling, intelligent power management, adaptive power management, dynamic bandwidth switching, and other forms of variable power and performance functionality for servers, storage, networks, and even facilities cooling. Technologies are being integrated from the design up, that is, from the lowest-layer silicon and gates on a chip for power management to vary performance and energy use as well as enhanced chip-level cooling.

Chip-level cooling is evolving from add-on precision heat-removing technology to cooling embedded inside the heat-producing silicon. In addition to removing heat from inside as well as outside chips and other components, the actual chips themselves, or the silicon they are made of, are being adapted to consume less power while doing more work. When they are not busy, the chips are being enhanced with the ability to turn certain gates or portions of the chip off, to reduce power usage while remaining available to perform certain functions. Another variation is the ability to slow down and operate at a slower pace and consume less energy when the chip is not required to run at full speed. Chips are not the only area where improvements are appearing; enhancements to magnetic disk drives and other power-consuming devices are appearing as well. Power supplies found in servers, storage, and networking components are becoming more efficient, with some exceeding 90% efficiency.

Continued movement toward adding more processing power and heavy thinking capabilities into storage systems as well as external network or fabric-based appliances or blades will enable future abstraction and management functions. While early attempts at creating large pools of consolidated storage have occurred and more will take place, in the future a significant value for storage will be that virtualization will shift from a focus on consolation to abstraction, emulation, and management, similar to what is being done with servers. The value proposition will be to enable transparent movement and migration of data to support routine IRM tasks including upgrades, technology refresh, maintenance, business continuity/disaster recovery (BC/DR), and other tasks. This does not mean that storage functionality will go away or that the only place for intelligence and advanced functionality will be in the network. What it means is that network-based functionality that complements and leverages underlying storage systems capabilities will finally live up to its potential without adding cost, complexity, overhead, or instability.

The trend of decoupling or disaggregating hardware and software, applications and hardware, local and remote resources with increased abstraction of physical resources to logical applications and business function will continue. Buzzwords, including service-orientated architecture (SOA), software as a service (SaaS), cloud, grid, and MSP, among others, will continue to proliferate, as will their use as marketing vehicles. Cutting through the noise, however, there are and will be new value-added services to leverage as part of a larger virtual data center that encompasses various resources in different locations.

The objective is to lower cost, boost productivity, and improve service delivery to meet performance and availability objectives in a flexible or agile manner that also helps the environment. Cloud computing is the latest repackaging of services that in the past have been called application service providers (ASP), storage service providers (SSP), data hotels, service bureaus, and outsourcing, among other names. A benefit is that all or portions of a business's IT capabilities can be shifted to a cloud or outsourced model, but the business benefit may not be so obvious. A careful and balanced analysis looking at business, technology, and financial concerns may find the best approach to be a combination of in-house optimized service delivery with other outsourced or managed services, including BC/DR, backup and replication, email or online storage for mobile works, pilot applications to avoid near-term investment overhead, as well as hosting of mature, stable, and legacy applications.

General trends and technology improvements expected include:

- Faster processors that consume less power while doing more work in a smaller footprint
- Increased usage of SSD devices, taking pressure off disk drives
- Shift of focus for magnetic disk drives to take pressure off magnetic tape
- Magnetic tape focus shifts to storing densely compressed data for long-term, low-cost archiving, and data preservation of in-active, or idle data

- Expanding focus from data center to offices, including remote offices and home offices

- Converged I/O and networking to support converged servers and storage

- Continued focus on desktop, laptop, and workstation virtualization

- Energy Star compliance for data centers, servers, storage, and networks

- Environmentally friendly inks and recyclable paper for printers and copiers

- More awareness of the many faces or facets of being green

- Thermal management including intelligent, precision, and dynamic or smart cooling

- Changing data access patterns and life-cycles requiring more data to be accessible

- Less emphasis on moving data around, more emphasis on reducing data storage impact

- Metrics to reflect active work and data being stored along with idle-time energy-saving modes

- Alignment of metrics to application and business value and level of service being delivered

- A shift from focusing on consolidation to focusing on enabling growth and productivity

Keep in mind that the destination is not grids, clouds, SaaS, xSP converged or consolidated servers, storage and networks, or other technologies. Rather, these are vehicles to enable and facilitate change to meet various demands while delivering IT services in a cost-effective and environmentally friendly manner. Even with the promise of shifting to cloud, SaaS, managed service providers, grids, or other forms of virtualized and abstracted physical resources, someone somewhere will have to take care of the physical and virtual resources that are essential to support virtualized environments.

11.3   How We Can Get There—Best Practices and Tips

Use different tiers of technology to address various needs and issues, depending on cost and service requirements. This is similar to a common approach to managing energy and the environment in general. Simply building more power plants is similar to building more data centers; however, both require time, money, approvals, and various other resources. Diversity and leveraging of tiered resources is an alternative approach that aligns the right resource to the task at hand, leveraging best practices and managing use and productivity.

General tips on how to address PCFE and related issues as well as improve the efficiency and productivity of IT data centers include the following:

- Where practical and possible, power down equipment or enable power-saving modes.
- Adopt tiered data protection across applications, servers, storage, and networks.
- Eliminate heat as close to the source as efficiently and safely possible.
- Leverage virtualization aggregation, emulation, and management abstraction capabilities.
- Reduce data footprint impacts by gaining control and managing data and storage.
- Align tiered servers, storage, networks, and facilities to specific service-level needs.
- Avoid performance and availability bottlenecks from overconsolidating resources.
- Use energy-efficient technologies that do more work or store more data per unit of energy used.
- Review facilities and IT equipment power usage and cooling efficiency.
- Explore intelligent, smart, and precision cooling and thermal management technology.

- Measure, monitor, and manage resource usage relative to business productivity and activity.

- Coordinate capacity planning across servers, storage, networks. and facilities.

- Eliminate halon and other hazardous substances and materials from facilities.

- Investigate emerging technologies, including thin provisioning and data de-duplication.

- Utilize alternative energy sources and cooling options, including economizers.

- Assess suppliers and partners as part of a green ecosystem and supply chain.

- Redeem energy rebates, discounts, and other financial incentives.

11.4   Chapter and Book Summary

There are many different facets to being green that require solutions. So, too, are there many different technology solutions looking for problems. Environmental concerns can be addressed by applying efficiency and optimization to operate more economically. Consider different technologies and approaches to using various tools in the context of the business problem or opportunity to be addressed. For example, although consolidation can bring great hardware and PCFE savings, consider also the subsequent performance impact on business productivity as well as savings from reducing software footprint.

Consolidation of physical resources is becoming understood, and it is relatively easy to see the results. However, because hardware costs represent only a small fraction of the total cost to deliver an IT service, additional savings and benefits can result from addressing infrastructure resource management and software footprint impacts. For example, if 10 physical 2,000-W servers along with their associated operating systems and applications are consolidated to a single server, that represents an energy reduction of 18 kWh or the equivalent of 360 fifty-watt light bulbs. However, despite the reduction in hardware as well as PCFE impact, there are still 10 operating systems and their applications, along with 10 backups and other IRM tasks to contend with. Future opportunities will be to consolidate applications to reduce the number of operating systems or maintain the existing operating system footprint count to support growth.

Look at technology adoption in the context of market or industry versus actual customer deployment. Density brings the benefits of a reduced physical footprint but also increased power, cooling, weight, and management footprints. Server or storage consolidation in general reduces the number of physical units to be managed, but server and storage virtualization in their current forms do not help to aggregate or consolidate the number of operating systems or application instances. Avoid simply moving problems from one area to another, for example, by solving a server-related problem that, in turn, causes a data storage, network, or backup problem.

Moving forward, as tools and technology emerge and mature, significant gains can be achieved by reducing the footprint and impact of multiple operating systems and application instances along with associated software and maintenance costs. Although IT data centers, along with other habitats for technology, in general use only a small percentage of the total electricity used—contributing only about 1 to 2% to overall global emissions—the bigger green and environmental issue is the growing demand and reliance on information being available and economic issues including the green footprint of the associated IT industry green supply chain.

In most IT data centers, as well as in the tech industry at large, the focus is on application servers to reduce electric power consumption and cooling demands. However, given the anticipated continued growth in the amount of data generated, processed, and stored for longer periods of time, the focus will shift to optimizing data storage and associated networking components in the data center. In order to power IT equipment in the future, you may not be able to assume that your energy provider will be able to supply you with adequate and reliable power in a cost-effective manner.

Future and sustained ability to power IT equipment to enable business applications growth and enhancements hinges on being able to maximize current power and cooling capabilities supplemented by alternative power and cooling resources. This means combining different techniques to do more with what you have, reducing your data footprint, and maximizing the efficiency and effectiveness of IT equipment usage without negatively impacting application service levels.

There are Many different issues that require a balancing of resources. Manufacturers are enabling enterprise IT data centers of all sizes to

address power, cooling, floor space, and associated environmental challenges with various technologies to balance the supply of available power,
cooling, and floor space with demand-side needs of enabling business
applications in a scalable, flexible, and resilient manner. The bottom line
is that for most IT data centers there is no one single silver bullet. However, when various technologies are combined and balanced with best
practices, energy savings to support growth are achievable.

As more IT applications are virtualized, there will need to be an
increase in the understanding of how to maximize, leverage, and manage
infrastructure resources effectively, as opposed to simply throwing more
hardware at a problem. Increase your awareness of applicable issues as well
as what you can do today and tomorrow to support your specific needs.
This includes educating yourself on the issues and alternative approaches
and solutions as well as looking at power efficiency and effectiveness versus
simply avoiding power usage.

Shift your thinking from acquiring storage resources on a cost-per-
capacity basis to one of how much effective and useful storage with
advanced data protection capabilities can be acquired and operated in a
given footprint that is energy-efficient and helps to reduce and manage
your data footprint more effectively. If you have not already done so, align
data center facilities personnel and server, storage, and networking staff to
work together to understand each others' needs and requirements as well
as internal causes and effects.

Significant improvements in energy consumption and cooling capabilities can be achieved by combining various technologies and techniques to
address different issues. The key is to understand what approach or technology to use when and where to address which issue without introducing or
exacerbating other problems.

Principal action and takeaway points from this book include:

- Identify green- and PCFE-related issues and requirements, near-
  term and long-term.

- Establish near-term and long-term strategies and plans for deployment.

- Look beyond carbon footprint messaging to see how to address various PCFE issues.

- Align applicable technology and techniques to the task at hand.
- Balance performance, availability, capacity, and energy use for a given level of service.

Keep in mind:

- You can't go forward if you can't go back.
- You can't delete what you have not preserved—assuming data has some value.
- You can't preserve what you can't move—use manual or automated tools.
- You can't move what you don't manage—rules and policies are needed.
- You can't manage what you don't know about—you need insight into what exists.

Green washing and green hype may be on an endangered species list, but addressing core IT data center issues to enable more efficient and productive IT service delivery in an economical manner will also lead to environmental benefits. Addressing green and PCFE issues is a process; there is no one single solution or magic formula. Rather, a combination of technologies, techniques, and best practices to address various issues and requirements is needed. Green washing and green hype may fade away, but PCFE and related issues will not, so addressing them is essential to IT, business growth, and economic sustainment in an environmentally friendly manner.

About the Author

With a career spanning three decades in the IT industry, Greg Schulz is the founder of the StorageIO Group (www.storageio.com), a leading technology industry analysis and consulting firm focused on data infrastructure topics. StorageIO provides services to established and emerging technology vendors, value-added resellers, end users, the media, and venture capital organizations on a global basis.

Greg gained his diverse industry insight from being in the trenches of IT data centers. He has held numerous positions, including programmer, systems analyst, server and storage systems administrator, performance and capacity planning analyst, server and storage planner, and internal disaster recovery consultant, at diverse companies including photographic services, electricity generating and transmission utilities, financial services organizations, and transportation companies. Greg has also worked for various storage and networking companies providing hardware, software, and services solutions in a variety of roles ranging from systems engineering and sales to marketing and senior technologist.

Before founding the StorageIO Group in 2006, Greg was a senior analyst at an analysis firm covering SAN, NAS, and associated management tools, techniques, and technologies. He has been involved with industry organizations such as the Computer Measurement Group (CMG), the RAID Advisory Board (RAB), the Storage Networking Industry Association (SNIA), and the SNIA Green Storage Initiative (GSI), among others.

Greg has published extensively, appearing regularly in print and online. In addition to his reports, columns, articles, tips, podcasts, videos, and webcasts, Greg is the author of *Resilient Storage Networks—Designing Flexible Scalable Data Infrastructures* (Elsevier, 2004) and a co-author or contributor to other projects including *The Resilient Enterprise* (Veritas, 2002). He is also a frequent speaker on data infrastructure and related management topics at conferences and custom events around the world. Greg has an bachelor's degree in computer science and a master's degree in software engineering.

Appendix A

Where to Learn More

Companion materials and resources for further reading—including additional technology and solution examples, metrics, calculators, and other useful links—can be found at www.thegreenandvirtualdatacenter.com. Additional information about data and storage networking is available in the author's book, *Resilient Storage Networks—Designing Flexible Scalable Data Infrastructures* (Elsevier, 2004).

www.80plus.org	Energy-efficient power supply trade group
www.afcom.com	Data center industry user group
www.aiim.com	Archiving and records management trade group
www.ashare.org	HVAC Engineers Association
www.cdproject.net	Carbon Disclosure Project
www.climatesaverscomputing.com	Green computing industry trade group
www.cmg.org	Computer Measurement Group and capacity planners
www.cmp.com	CMP Media Group
www.computerworld.com	Publication covering servers, storage, and networking
www.dsircusa.org	Database of State Incentives and Renewable Energy
www.drj.com	Disaster Recovery Journal
www.echannelline.com	Channel-focused publication venue

www.eere.energy.gov	U.S. Department of Energy (DoE) website
www.eia.doe.gov/fuelelectric.html	Portal for electrical power generation, use, and costs
www.energyshop.com	Portal for energy pricing and options
www.energystar.gov	U.S. EPA Energy Star website
www.enterprisestorageforum.com	Publication coverage data and storage management
www.epa.gov/stateply/	U.S. EPA Climate Leaders' Initiative
www.epeat.net	Site for comparing desktop and related products
www.fcoe.com	Website pertaining to Fibre Channel over Ethernet
www.fibrechannel.org	Fibre Channel Industry Association website
www.fueleconomy.gov	U.S. government site for energy efficiency
www.greendatastorage.com	Site pertaining to green data storage and related topics
www.greenpeace.org	Greenpeace website
www.greenwashing.net	Information about green washing
www.ieee.org	Institute of Electrical and Electronics Engineers
www.ietf.org	Internet Engineering Task Force
www.iso.org	International Standards Organizations
www.jupitermedia.com	Jupiter Media Group, including Internet News
www.naspa.org	System administrators' user group
www.nfpa.org	National Fire Protection Association
www.pcisig.com	Peripheral Component Interconnect (PCI) trade group

www.processor.com	Publication covering servers, storage, and networking
www.sans.org	Security-related website
www.scsita.org	SCSI trade association
www.snia.org	Storage Networking Industry Association
www.spec.org	Server and storage benchmarking site
www.storageio.com	StorageIO Group website
www.storageioblog.com	Author's blog site
www.storageperformance.org	Storage performance benchmarking site
www.svlg.net	Silicon Valley Leadership Group
www.t11.org	Fibre Channel and related standards
www.techtarget.com	IT data center publications, conferences, and events
www.thegreengrid.org	Industry trade group
www.tiaonline.org	Telecomucaitons Industry Association
www.top500.org	List of top 500 supercomputing sites
www.tpc.org	Transaction Performance Council benchmark site
www.trustedcomputinggroup.org	Security-related website
www.uptimeinstitute.org	Uptime Institute
www.usenix.org	LISA and data center forums
www.usgbc.org	United States Green Building Council
www.communities.vmware.com	VMware technical community website
www.wwf.org	World Wildlife Fund
www.zjournal.com	Enterprise-focused servers, storage, and networking

In addition to the above links, information about industry and technology solution providers, trade groups, publications, and related information can be found at www.storageio.com/interestinglinks.html and www.greendatastorage.com/interestinglinks.html.

More details about the U.S. EPA Energy Star program and initiatives for data centers, servers, storage, networking, and related topics can be found at www.energystar.gov/index.cfm?c=prod_development.server_efficiency.

Microsoft Exchange Solution Reviewed Program (ESRP) benchmarking information can be found at http://technet.microsoft.com/en-us/exchange/bb412165.aspx.

Appendix B

Checklists and Tips

B.1 Facilities, Power, Cooling, Floor Space and Environmental Health and Safety

- Consolidate where practical without negatively impacting performance.

- Leverage faster servers, storage, and networks to replace multiple slower devices.

- Power down where practical; reduce power usage when possible.

- Review heating, ventilating, and air conditioning (HVAC) and energy usage, including thermal assessment.

- Adopt smart, adaptive, intelligent, and precision cooling technologies.

- Use economizers and other alternative cooling techniques where possible.

- Balance energy use between traditional and renewable energy sources.

- Address cooling air leaks, including gaps in raised floor tiles.

- Eliminate halon and other hazardous substances.

- Gain insight as to how energy is used to support business and Infrastructure resource management (IRM) functions.

B.2 Variable Energy Use for Servers

- Establish power management policies to balance productivity and energy savings.

- Enable power management settings on servers, desktops, and work-stations.

- Reduce power and performance when possible.

- Use standby, hibernate, and sleep modes when practical.

- Upgrade operating systems and third-party power management software.

- Implement smart power management and Energy Star servers.

- Consider how much work can be done per watt of energy when servers are active.

- Consider how much energy can be saved when servers are inactive, by using low-power modes.

- Look for adaptive power management features in new servers.

- New servers should feature 80%-plus efficient power supplies and cooling fans.

B.3 Variable Energy Use for Storage

- Establish power management policies to balance productivity and energy savings.

- Not all storage for all applications lend themselves to being powered down.

- Power-down modes are applicable to offline and some near-line applications.

- Power-saving modes can reduce power while keeping disks and stor-age online.

- Are specific disk drives required for use with MAID (Massive Array of Idle Disks) functionality?

- What are the performance impacts of enabling MAID?

- Are drives powered down completely or put into power-saving modes?

- How many power-saving modes or different performance levels are available?

- What maintenance features are available to safeguard disk drives that are powered down?

- What is the impact to file systems when a LUN (Logical Unit Number) and associated disks are powered down?
- What granularity is enabled or available with the MAID or intelligent power management (IPM) storage?
    - Is MAID enabled across the entire storage system?
    - Is MAID enabled on a volume or RAID group basis?
    - Is MAID enabled across different RAID (Redundant Array of Independent Disks) levels?
    - Is MAID enabled across different types of disk drives?

B.4   Data Footprint Impact Reduction

- Identify orphaned storage that is allocated but not being used.
- Keep performance and availability in perspective when reducing data footprint.
- Develop and implement a comprehensive data footprint reduction strategy:
    - Archive inactive data to offline or removable media.
    - Delete data that is no longer needed or required.
    - Leverage real-time compression of online active and primary storage.
    - Utilize compression for offline and removable media for backup and archiving.
    - Apply de-duplication technology to recurring or frequently accessed data.
    - Balance high compression ratios for density with compression rates to move data.
    - Implement space-saving snapshots and data movement techniques.
- Items and caveats to keep in mind include:
    - Impact on active and online applications and data from reducing data footprints
    - Fragmentation of storage from de-duplicated or reduced data footprints
    - Encryption coexistence with de-duplication and compression technologies

- Impact and overhead of inline de-duplication compared to postprocessing
- How much buffer or temporary storage is needed for performing de-duplication
- Regulatory and compliance requirements when using de-duplication
- How much free space needs to be maintained for solutions to function properly
- Whether solutions enable mixed modes, on-the-fly or real-time, including in-line, in-band, and immediate as well as deferred or postprocessing, out-of-band, or scheduled de-duplication

B.5   Security and Data Protection

- Know where your data and IT assets are, where data has been, and who has access to it.
- Consider and keep in perspective internal as well as external threat risks.
- Encrypt all data on portable and removable media (disk, tape, optical).
- Encrypt data in flight as well as at rest.
- To avoid being circumvented, find ways to prevent security from affecting productivity.
- Implement tiered data protection and security with multiple layers of defense.
- Assign descriptors or monikers to identify which tapes require which key.
- Assign different keys to different tapes or groups of tapes.
- Utilize 128- and 256-bit AES encryption capabilities.
- Utilize tamper proof access and audit trail logs.
- Ensure secure shredding of encrypted data.
- Use flexible and easy-to-use key creation, assignment, and escrow protocols.

- Ensure coexistence with other key management and encryption products.
- Use high performance encryption to avoid bottlenecks.

B.6 How Will Virtualization Fit into Your Existing Environment?

- What are your various application requirements and needs?
- Will you use virtualization for consolidation or facilitating IT resource management?
- What other technologies do you currently have or plan to have?
- What are your scaling (performance, capacity, availability) needs?
- Who will deploy, maintain, and manage your solutions?
- Will you be shifting the point of vendor lock-in or increasing costs?
- What are some alternative and applicable approaches?
- Do you need virtualization, or do you want virtualization technology?
- How will a solution scale with stability?

B.7 Desktop, Remote Office/Branch Office (ROBO), Small/Medium Business (SMB), and Small Office/Home Office (SOHO) Users

- Enable bandwidth or application optimization (WAFS, WAAS, or WADM) technologies.
- Implement applicable security to enable productivity while protecting IT resources.
- Gain management control of remote and distributed data before relocating data.
- Ensure that remote and mobile data is backed up and synchronized for data protection.
- Investigate the applicability of desktop virtualization to enable simplified management.

- Implement encrypted removable media and enable encryption on mobile devices.
- Re-architect backup and data protection for desktop and remote IT data and resources.
- Investigate and adopt environmentally safe inks, toners, and paper for printers.
- Safely dispose of retired or surplus IT equipment and media.
- Leverage Energy Star desktop and laptop computers as well as printers and copiers.
- Enable energy-saving modes on computers and turn monitors off when not in use.
- Learn more about Energy Star support programs and utilizes at:
  - www.energystar.gov/index.cfm?c=power_mgt.pr_power_mgt_users

B.8   Questions to Ask Vendors or Solution Providers

- What is the vendor's green message, and how does it align to your specific needs?
- What solution offerings does the vendor have to address power, cooling, floor space, and environmental as well as other green issues?
- What are the directions and roadmaps for future initiatives and enhancements?
- Look for a comprehensive green message encompassing:
  - A website that details how a company is green, rather than merely talking about being green
  - How business is conducted, including corporate environmental health and safety (EHS) as well as other green programs
  - A green supply chain management involving partners and suppliers
  - What awards have been achieved along with regulatory compliance
  - Material safety data sheets (MSDS) and statements of EHS compliance

- Recycling, reuse, and reduction programs and accomplishments
- Active participation in industry trade groups and associated initiatives
- Involvement in initiatives such as the Carbon Disclosure Project and Climate Leaders
- Product and technology disposition

B.9   General Checklist and Tip Items

- Eliminate hazardous substances, including those currently exempted under regulations governing removal of hazardous substances.
- Implement a green supply chain and EHS management system.
- Look at shipping functionality compared to statement of direction (SOD) or futures.
- Balance futures with what works today—leverage what works, learn about the trends.
- Enhance your awareness of your own energy footprint.
- Review your energy bill or statement to see how much energy is being used and the cost.
- Have a power, cooling, and thermal assessment performed for facilities.
- Explore raising temperatures in facilities while staying within vendors' guidelines.
- Balance performance, availability, capacity, and energy (PACE) to service levels.
- Have a strategy—execute it, monitor the results, and revise it as needed.
- Implement meaningful measurements that reflect work performed and energy saved:
  - Compare energy used for performing useful work and storing active data
  - Compare energy used for inactive and energy-saving modes or idle storage

- Turn off video monitors and lights when they are not in use.
- Balance energy avoidance with energy efficiency to support productivity and cost savings.
- Establish data retention and archiving policies and management.
- Establish a formal green policy for energy efficiency, EHS, cooling, and emissions.
- Monitor, measure, manage, and revise as a continuum.
- Be aware of the costs of going green:
  - EHS and technology disposal and recycling costs.
  - Costs of new hardware, software, and services along with maintenance fees.
  - Data management tools require software and services.
  - Software requires hardware that consumes power.
  - Moving data across different tiers of storage requires energy.
  - Performance can be impacted in the quest to reduce data footprints.
  - Balance performance, availability, capacity, and energy to quality-of-service level.
  - Virtualization improves utilization but not necessarily performance.
  - Provisioning tools mask allocation, not actual usage.

Glossary

Table C.1   IRM Counting and Number Schemes for Servers, Storage, and Networks

	Binary Number of Bytes	Decimal Number of Bytes	Abbreviations
Kilo	1,024	1,000	K, ki, kibi
Mega	1,048,576	1,000,000	M, Mi, bebi
Giga	1,073,741,824	1,000,000,000	G, Gi, gibi
Tera	1,099,511,627,776	1,000,000,000,000	T, Ti, tebi
Peta	1,125,899,906,842,620	1,000,000,000,000,000	P, Pi, pebi
Exa	1,152,921,504,606,850,000	1,000,000,000,000,000,000	E, Ei, exbi
Zetta	1,180,591,620,717,410,000,000	1,000,000,000,000,000,000,000	Z, Zi, zebi
Yotta	1,208,925,819,614,630,000,000,000	1,000,000,000,000,000,000,000,000	Y, Ui, yobi

Additional glossaries and dictionaries can be found on various industry trade group, as well as vendors', websites, including those listed in Appendix A, and on the links page of this book's companion website: www.thegreenandvirtualdatacenter.com.

100GbE	100-Gigabit Ethernet
10GbE	10-Gigabit Ethernet
16GFC	16-Gigabit Fibre Channel
1GbE	1 Gigabit Ethernet
1GFC	1-Gigabit Fibre Channel
24x7	24 hours a day, 7 days a week; always available
2GFC	2-Gigabit Fibre Channel

3G	Third-generation cellular broadband communications
3R	Recycle, Reduce, Reuse
40GbE	40-Gigabit Ethernet
4GFC	4-Gigabit Fibre Channel
4R	Recover, Restore, Rollback, Restart or Resume
8GFC	8-Gigabit Fibre Channel
AC	Alternating current electricity
ACPI	Advanced configuration and power interface
AES	Form of encryption
AFR	Annual failure rate measured or estimated failures per year
Agent	Software for performing backup or other infrastructure resource management functions on a server
AHU	Air-handling unit for cooling and HVAC systems
Air scrubber	Device for removing contaminants or emissions from air
AMD	Manufacturer of processing chips
ANSI	American National Standards Institute
Antimony	Hazardous substance found in some computer related devices
API	Application Program Interface
APM	Adaptive power management; varies energy use to service delivered
Application blades	Server, storage, and input/output networking blades for various applications
Applications	Programs or software that performance business or infrastructure resource management services
Archiving	Identifying and moving inactive data to alternate media for future use

ASP	Application service provider delivering functionality via the Internet
Asynchronous	Time-delayed data transmission used for low-cost, long-distance transmission
ATM	Asynchronous Transfer Mode networking technology
Availability	The amount or percent of time a system is able and ready to work
AVS	Adaptive Voltage Scaling; varies energy used to work performed
AVSO	Adaptive voltage scaling optimized, or enhanced energy savings
Bandwidth	Measure of how much data is moved in a given amount of time
BC	Business continuance
B2D	Backup to disk
Blade center	Packaging combining blade servers, input/output, and networking blades
Blade server	Server blade packaged as a blade for use in a blade center
Boot	Process of starting up and loading software into a server
Br	Bromine; a hazardous substance found in IT equipment
BTU	British thermal unit; a measure of amount of heat
CapEx	Capital expenses
Carbon emissions	Carbon dioxide (CO_2) emissions as a by-product of electricity generation
Carbon offset credits	Means of offsetting carbon emissions by paying or buying credits
Carbon tax	Tax on carbon emissions that result from energy use
CAS	Content addressable storage

CD	Compact disc
CDP	Continuous data protection or complete data protection
CEE	Consortium for Energy Efficiency, or Converged Enhanced Ethernet
Chiller	Device for removing heat from coolant
Cl	Chlorine, a substance used in some electronic equipment
CIFS	Common Internet File system (NAS) for file and data sharing
CIM	Common information model for accessing information
Citix	Virtualization solutions provider
Cloud computing	Internet or Web-based remote application or infrastructure resource management–related services
Cluster	Collection of servers or storage working together, also known as a grid
Clustered file system	Distributed file system across multiple servers or storage nodes
CMDB	Configuration Management Database or repository
CMG	Computer Measurement Group; capacity and performance planners
CNA	Converged Network Architecture
CNIC	Converged Network Interface Card or Chip
CO_2 emissions	Carbon dioxide emissions
Cold aisles	Aisles in between equipment racks or cabinets supplied with cold air
Consoles	Management interfaces for configuration or control of IT devices
Cooked storage	Formatted, usable storage with a file system or non-raw storage

Cooling ton	12,000 BTU to cool 1 ton of air
COS	Console operating system, also known as a console or boot system
CP	Capacity planning
CPU	Central processing unit
CRAC	Computer room air conditioning
Cross technology domain	Solution or tools that address multiple technologies and disciplines
CSV	Comma Separated Variable format used for spreadsheet data
D2D	Disk-to-disk snapshot, backup, copy, replication, or archive
D2D2D	Disk-to-disk-to-disk snapshot, backup, copy, replication, or archive
D2D2T	Disk-to-disk-to-tape snapshot, backup, copy, replication, or archive
D2T	Disk-to-tape backup, copy, or archive
DAS	Direct Attached Storage, either internal or external to a server
Data barn	Large repository for holding large amounts of online or offline data
Data in flight	Data being moved between locations, between servers or storage
Database	Structured means of organizing and storing data
DB2/UDB	IBM database software
DBA	Database administrator
DBS	Dynamic bandwidth switching, varies energy use to performance
DC	Data center or direct current electricity
DCE	Data center Ethernet for converged input/output and networking

DCE	Data communications equipment or distributed computing environment
DCiE	Data center infrastructure efficiency
DCPE	Data center performance efficiency
DDR/RAM	Double data rate random access memory
De-dupe	De-duplication or elimination of duplicate data
DEFRA	UK Department for Environment, Food, Rural Affairs
Desktop	Workstation or laptop computer, also known as a PC
DFS	Distributed file systems for distributed and shared data access
DHCP	Dynamic Host Configuration Protocol for network management
DIO	Direct input/output operations addressing specific storage addresses
Director	Input/output and networking large-scale, multi-protocol, resilient switch
DL	Disk library used for storing backup and other data, alternative to tape
DLM	Data life-cycle management
DMA	Direct memory access
DMTF	Distributed Management Task Force
DNS	Domain name system for managing Internet domain names
DOE FEMP	U.S. Department of Energy Federal Energy Management Program
DoE	U.S. Department of Energy
DPM	Data protection management
DR	Disaster recovery
DRO	Data replication optimization
DRP	Disaster recovery planning

DSM	Demand-side management
DVD	Digital video disc
DVR	Digital video recorder
DWDM	Dense wave division multiplexing
ECC	Error correcting code
ECCJ	Energy Conservation Center Japan
Economizer	Cooling device that utilizes outside cool air for cooling
eDiscovery	Electronic search and data discovery
EHS	Environmental health and safety
ELV	End of Life Vehicle Directive—European Union
Emissions tax	Tax for emissions such as CO_2 from energy generation or use
EMS	Environmental management system
Energy Star	U.S. EPA Energy Star program
EPA	U.S. Environmental Protection Agency
EPEAT	Electronic Product Environmental Assessment Tool
ESRP	Microsoft Exchange Solution Reviewed Program
Ethernet	Network interface
ETS	Emissions trading scheme
EU	European Union
E-waste	Electronic waste associated with disposal of IT equipment
FAN	File area network or file-based storage management
FC	Fibre Channel
FCIA	Fibre Channel Industry Association
FCIP	Fibre Channel on IP for long-distance data movement and mirroring
FCoE	Fibre Channel over Ethernet

FCP	Fibre Channel SCSI Protocol
FC-SB2	FICON Upper Level Protocol (ULP)
FEMP	U.S. Department of Energy Federal Energy Management Program
File data access	Accessing data via a file system locally or remotely
Firewall	Security device or software to block unauthorized network access
FLASH memory	Nonvolatile memory
FTP	File Transfer Protocol
Fuel cell	Alternative source for producing energy including electricity
G&T	Generating and transmission network for electrical power
GbE	Gigabit Ethernet
Generator	Device for producing electrical power for standby or co-generation
GHG	Greenhouse gas
Ghz	Gigahertz frequency, a measure of speed
Global namespace	Directory name space to ease access across multiple file systems
GPS	Global Positioning System (or initials of the author of this book)
Green supply chain	Suppliers and partners adopt and implement being green
Green technology	Technology that addresses power, cooling, floor space, or environmental health and safety
Green washing	Using green to market or being seen as green rather than actually being green
Grid	Local or wide area cluster of servers or storage working together
Guest OS	Guest operating system in a virtual machine or logical partition (LPAR), also known as an image

GUI	Graphical user interface
HA	High availability
HBA	Host Bus Adapter for attaching peripherals to servers or storage
HCA	Host channel adapter for InfiniBand
HD	High-definition broadcast or video
HDD	Hard disk drive such as Fibre Channel, SAS, SATA, or USB
HDTV	High-definition TV
HFC	Hydrofluorocarbons, compounds used in computer-related devices
HHDD	Hybrid hard disk drive with RAM, FLASH, and/or magnetic media
Hosting site	Facility or service provider that hosts IT components and services
Hot aisles	Aisles in between equipment cabinets where warm air exhausts
HPC	High-performance computing
HSM	Hierarchical storage management
HTTP	Hypertext Transfer Protocol for serving and accessing Web pages
HVAC	Heating, ventilating, and air conditioning
HyperV	Microsoft virtualization infrastructure software
Hypervisor	Virtualization framework that emulates and partitions physical resources
I/O rate	How many input/output operations (IOPS) read or write in a given time frame
I/O size	How big the input/output operations are
I/O type	Read, write, random, or sequential
I/O	Input/output operation, read or write
IBA	InfiniBand architecture
IC	Integrated circuit

IEEE	Institute of Electrical and Electronic Engineers
IETF	Internet Engineering Task Force
ILM	Information life-cycle management
IM	Instant messaging
Image	Guest operating system or workload residing in a virtual machine or logical partition (LPAR)
IMPI	Intelligent Platform Management Interface
Intel	Large processor and chip manufacturer
Iometer	Load generation and simulation tool for benchmark comparisons
IOPS	Input/output operations per second for reads and writes of various sizes
Iostat	Input/output monitoring tool
IOV	Input/output virtualization including converged networks and Peripheral Computer Interconnect (PCI) switching
IP	Intellectual property or Internet Protocol (part of TCP/IP)
IPM	Intelligent power management; varies energy used to service delivered
IPSec	Internet Protocol (IP)–based security and encryption
IPTV	Internet Protocol (IP)–based TV
IRM	Infrastructure resource management
iSCSI	SCSI command set mapped to Internet Protocol (IP)
ISO 14001	Environmental management standards
ISO	International Standards Organization
ISV	Independent software vendor
IT	Information technology
JEDEC	Joint Electronic Device Engineering Council

JEITA	Japan Electronics Information Technology Industry Association
J-MOSS	Japan version of removal of hazardous substances (RoHS)
Joule	Measure of energy usage
JRE	Java Runtime Environment
JVM	Java Virtual Machine
Key management	Managing encryption keys
Kill a Watt by P3	Device for monitoring watts, volts, and amperes
KPI	Key performance indicator
KVM	Keyboard video monitor
Kyoto Protocol	Multinational protocol and treaty to reduce emissions of greenhouse gases
LAN	Local area network
Laptop	Portable computer
LEED	Leadership in Energy Efficiency Design
Linux	Open-source operating system
LiveMigration	Virtual Iron function, similar to VMware VMotion
LPAR	Logical partition or virtual machine
LPG	Liquefied propane gas
LUN	Logical Unit Number addressing for storage targets or devices
MAC	Media Access Control layer for networking interfaces
Magnetic tape	Low-cost, energy-efficient, removable medium for storing data
MAID	Massive Array of Idle Disks, which avoids using power when not in use
MAID 2.0	Second-generation MAID with intelligent power management

Mainframe	IBM legacy large server; generic name for a large frame-based server
MAN	Metropolitan area network
Metadata	Data describing other data, including how and when it was used
MHz	Megahertz frequency, an indicator or speed
MIB	Management Information Block for Simple Network Management Protocol (SNMP)
MO	Magneto-optical storage medium
MPLS	Multi-Protocol Labeling Switching, a wide area network (WAN) networking protocol
MR-IOV	Peripheral Computer Interconnect (PCI) multi-root input/output virtualization (IOV) capability
MSDS	Material Safety Data Sheet for products
MSP	Managed service provider
MTBF	Mean time between failures—measured or estimated reliability
MTTR	Mean time to repair or replace a failed item
MW	Megawatts, a measure of power
NAND	Nonvolatile computer memory, such as FLASH
NAS	Network attached storage
Near-line	Nonprimary active data storage that does not need fast access
NEMA	National Electronic Manufactures Association
NERC	North American Electric Reliability Association
NFS	Network File System (network attached storage) for file and data sharing
Nfstat	Operating system utility for monitoring Network File System (NFS) activity
NIC	Network interface card or chip
NIST	National Institute of Standards and Technology
NOCC	Network operations control center

NPIDV	N_Port ID Virtualization for Fibre Channel input/output networking
NPVID	N_Port Virtual ID, similar to NPIDV
NRDC	Natural Resources Defense Council
NVRAM	Nonvolatile random access memory
Object data access	Data access via application specific Application Program Interface (API) or descriptors
OC	Optical carrier network
Offline	Data or IT resources that are not online and ready for use
OLTP	Online transaction processing
Online	Data and IT resources that are online, active and ready for use
OpEx	Operational expenses
Optical	Optical-based networking or optical-based storage medium
Orphaned storage	Lost, misplaced, or forgotten storage or storage space
OS	Computer operating system, such as Linux, UNIX, or Microsoft Windows
Outage	Systems or subsystems are not available for use or to perform work
Oversubscription	Allocating common shared service to multiple users to reduce costs
P2V	Physical-to-virtual migration or conversion of a server
PACE	Performance, availability, capacity, and energy
Para-virtualization	Optimized virtualization requiring custom software change.
Parity	Technique using extra memory or storage to ensure that data is intact

PATA	Parallel Advanced Technology Attachment (ATA) input/output interface
PC	Payment card industry for credit card security, or personal computer, or program counter
PCFE	Power, cooling, floor space, and environment
PCI	Peripheral Computer Interconnect for attaching devices to servers
PCIe	PCI express, the latest implementation of the Peripheral Computer Interconnect (PCI) standard
PCI IOV	Peripheral Computer Interconnect (PCI) Sig input/output virtualization implementation
PDA	Personal digital assistant
PDU	Power distribution unit
Pf	Power factor, how efficiently a power supply utilizes power
PG&E	Pacific Gas & Electric, utility company in Northern California
Physical volume	A disk drive or group of disk drives presented by a storage system
PIT	Point in time
PM	Physical machine or a real physical server or computer
PMDB	Performance management database
pNFS	Parallel Network File System (network attached storage) for parallel high-performance file access
POTS	"Plain old telephone system"
Power plant	Electric power-generation facility
Precision cooling	Pinpointing cooling closest to heat sources for efficiency
Primary	Storage, servers, or networks used day to day for service delivery

Provisioning	Allocating and assigning servers, storage, and networking resources
Proxy	Server configured to off-load servers for certain infrastructure resource management functions
PS	Power supply
PST	Microsoft Exchange email personal storage folder file
PUE	Power usage effectiveness
PVC	Polyvinyl chloride, a substance used in various products
QA	Quality assurance
QoS	Quality of service
Quad Core	Processor chip with four-core CPUs
RAID	Redundant Array of Independent Disks
Raised floor	Elevated floor with cabling and cooling located under the floor
RAM	Random-access memory
RASM	Reliability availability serviceability management
Raw storage	Storage that is not configured or formatted with a file system or RAID (Redundant Array of Independent Disks)
RDM	Raw device-mapped storage, as opposed to file-mapped storage
REACH	Registration, evaluation, authorization, and restriction of chemicals
REC	Renewable energy credit
Reliability	Systems function as expected, when expected, with confidence
Remote mirroring	Replicating or mirroring data to a remote location for business continuity and disaster recovery
Renewable energy	Energy sources such as wind, solar, or hydro

Replication	Mirroring of data to a second system or alternate location
RFID	Radio frequency identification tag and reader
RHDD	Removal hard disk drive
ROBO	Remote office/branch office
RoHS	Restriction of hazardous substances
ROI	Return on investment
Router	Networking or storage device for protocol conversion and routing
RPC	Remote Procedure Call for program-to-program communications
RPO	Recovery-point objective—the point to which data can be restored
RTO	Recovery-time objective—when recovered data will be usable again
RTSP	Real-Time Streaming Protocol for streaming data
RU or U	Rack unit
RUT	Rule of thumb
S/390	IBM mainframe architecture now referred to as "Z" or "Zed" series
SaaS	Software as a service
SAN	Storage area network
SAR	System analysis and reporting tool
SAS	Serial attached SCSI input/output interface and type of disk drive
SAS	Statistical analysis software
SATA	Serial Advanced Technology Attachment (ATA) input/output interface and type of disk drive
SB20/50	California Electronics Waste Recycling Act of 2003
SCADA	Supervisory control and data acquisition

Scheduled downtime	Planned downtime for maintenance, replacement, and upgrades
SCSI	Small Computer Storage Interconnect input/output interface and protocol
SCSI_FCP	Small Computer Storage Interconnect command set mapped to Fibre Channel, also known as FCP
SDK	Software development kit
Semistructured data	Email data that has structured or index and unstructured attachments
SFF	Small-form-factor disk drives, servers, input/output, and networking blades
SFP	Small-form-factor optical transceiver
SharePoint	Microsoft software for managing documents utilizing SQLserver
SIS	Single-instance storage, also known as de-duplicated or normalized storage
SLA	Service-level agreement, defining service availability and performance
SLO	Service-level objective in managing service delivery
SMB	Small/medium business
SMIS	Storage Management Interface Specification
Snapshot	A picture or image of the data as of a point in time
SNIA	Storage Networking Industry Association
SNMP	Simple Network Management Protocol for device management
SoA	Service-oriented architecture
SOHO	Small office/home office
SONET/SDH	Synchronous optical networking/synchronous digital hierarchy
SPC	Storage Performance Council

SPEC	Performance benchmarks
SQL database	Structure Query Language database;Microsoft database product that supports SharePoint
SR-IOV	Single-root PCIe input/output virtualization
SRM	Server, storage, or system resource management; or VMware site recovery manager for data protection management
SSD	Solid-state disk device using FLASH, RAM, or a combination
SSP	Storage solution provider, also known as managed service provider or cloud storage
Structured data	Data stored in databases or other well-defined repositories
Super-computer	Very fast and large performance-oriented server
SUT	System under test
SUV	System under validation
Switch	Input/output and networking connectivity for attaching multiple devices
Synchronous	Communications based on real-time data movement
T11	ANSI standards group for Fibre Channel
Tape	Magnetic tape used for storing data offline
TCP/IP	Transmission Control Protocol/Internet Protocol; networking protocols
Thin provisioning	Virtually allocate or overbook physical storage to multiple servers
Thumb drive	FLASH memory-based devices with USB interface for moving data
Tiered access	Different input/output and network interfaces aligned to various service needs
Tiered protection	Different data protection techniques and recovery-time objectives/recovery-point objectives for service needs

Tiered servers	Different types of servers aligned to various cost and service needs
Tiered storage	Different types of storage aligned to various cost and service needs
TPC	Transaction Processing Council
Transaction integrity	Ensuring write order consistency of time-based transactions or events
UK	United Kingdom
ULP	Upper-level protocol
UltraSCSI	Most recent form of parallel SCSI cabling and SCSI command set
UNIX	Open-source operating system
Unscheduled	Unplanned downtime for emergency repair or maintenance
Unstructured data	Data, including files, videos, photos, and slides, that is stored outside of databases
UPS	Uninterruptible power supply
Usable storage	Amount of storage that can actually be used when formatted
USB	Universal Serial Bus for attaching peripherals to workstations
USGBC	U.S. Green Building Council
V2P	Virtual-to-physical migration or conversion
V2V	Virtual-to-virtual migration or conversion of a server and applications
VA	Volt-amperes
VCB	VMware Consolidated Backup proxy-based backup for virtual machines
VDC	Virtual data center
VDI	Virtual desktop infrastructure
VIO	Virtual input/output
Virltual Iron	Virtualization infrastructure solution provider

Virtual memory	Operating system or virtual machine extended memory mapped to disk storage
Virtual office	Remote or home office for mobile or remote workers
Virtualization	Abstraction, emulation, and aggregation of IT resources
VLAN	Virtual local area network
VM	Virtual machine or logical partition that emulates a physical machine
VMark	VMware benchmark and comparison utility
VMDK	VMware disk file containing the virtual machine instance
VMFS	VMware file system, stored in a VMDK file
VMotion	VMware tool for migrating a running virtual machine to another physical server
VMware	Virtualization infrastructure solution
VOD	Video on demand
Volume manager	Software that aggregates and abstracts storage for file systems
VPN	Virtual private network
VTL	Virtual tape library based on disk drives emulating tape drives
VTS	Virtual tape system; same as a virtual tape library
WAAS	Wide area application services, similar to wide area file services (WAFS)
WADM	Wide area data management, similar to wide area file services (WAFS)
WADS	Wide area data services, similar to wide area file services (WAFS)
WAFS	Wide area file services, tools for remote data and application access
WAN	Wide area network

Web 2.0	Second-generation Web applications that are two-way or collaborative
WEEE	Waste from electrical and electronic equipment
WGBC	World Green Building Council
WHDI	Wireless high-definition transmission
Wide area cluster	Server or storage cluster, also known as grid, spread over a wide area
WiFi	Wireless networking over relatively short distances
WiMax	Higher-speed, longer-distance next-generation wireless networking
Windows	Microsoft operating system
Workstation	Desktop PC or laptop computer
WWPN	World Wide Port Name, used for power Fibre Channel addressing
x86	Popular hardware instruction set architecture designed by Intel
Xenmotion	VM movement utility for Xen, similar to VMware VMotion
XML	Extensible Markup Language
xWDM	Generic term for dense wave division multiplexing
Y2K	Millennium or year 2000
Zen	Open-source virtualization infrastructure used by Virtual Iron
zSeries	IBM legacy mainframe that supports zOS and open Linux and logical partitions (LPARs)

ndex

Numerics

Wh, 39
kWh, 39
0 kWh, 39
-Gb Ethernet, 181
-Gb Fibre Channel, 213
GbE iSCSI, 263
GbE-based iSCSI, 213
-Gbit Ethernet, 261, 279
94 Fire wire, 181
Gb Ethernet, 181
x7, 345
-inch Fibre Channel disk, 316
, 174
U cabinet, 159
Gb Fibre Channel, 187
server, 159
Plus Initiative, 182
Gb Fibre Channel, 187, 213
Gbit Fibre Channel (8GFC), 262
1 emergency, 4

A

straction, 188
cess Patterns, 210
uisition cost, 9
ivity-per-watt, 58
ptive power management, 7, 40
ptive voltage scaling (AVS), 224
S, 340, 346
ent-Based, 78

Agent-Based Backup, 79
Agent-Less Data Protection, 78
aggregating storage, 243
Aggregation, 54, 188
aggressive data deletion, 312
air, 161
air temperature, 161
Alternate Power Options, 136
Alternative Cooling, 142
alternative energy, 46
Alternative Energy Options, 136
Alternative Energy Sources, 41
AMD-V, 176
annual kWh, 38
appliance, 174
application programming interfaces
    (APIs), 71
application service providers (ASP), 326
Application Virtual Machines, 195
application-aware DPM, 85
Archiving, 40, 82
Arrays of Independent Disks (RAID),
    236
Asset management, 199
ATA, 213

B

Basic Input/Output System (BIOS), 173
BC/DR, 67, 302
Blade Center, 279
blade center servers, 62

Blade Centers, 163, 174, 184
Blade Servers, 184
Blade servers, 163, 202
BLEVE, 148
Boost energy efficiency, 301
British thermal units (Btu), 11
Btu, 38
BTUs, 298
business continuance and disaster recovery (BC/DR), 205
Business continuity (BC), 56
business continuity and disaster recovery (BC/DR), 67
business continuity, and disaster recovery (BC/DR), 196
business continuity/disaster recovery (BC/DR), 325

C

Cabinets, 182
Cable Management, 291
Capacity Planning, 25, 304, 321
Carbon credits, 13, 18
Carbon Disclosure Project, 321
carbon emissions taxes, 12
carbon footprint, 29
carbon monoxide, 145
carbon neutral, 13
carbon offset credits, 13, 310
Carbon offsets, 41
Central Processing Units, 175
central processing units (CPUs), 173
Chicago Climate Exchange, 13
Chip-level cooling, 325
CIFS, 269
Classes of Fires, 147
Closed cooling systems, 143
Cloud Computing, 155
Clustered and Bulk Storage, 216
clustering, 41, 50
clusters, 200
Clusters and Grids, 200

CO2 emissions, 8
Coal, 11
coal power plants, 11
COBOL, 168
Common Internet Format (CIFS), 25
Comparing Server Performance, 183
Complete Data Protection, 83
Compression, 230
computation-intensive, 8
computer room air conditioning (CRAC), 134
configuration management database (CMDB), 292
connectivity, 251
Consolidation, 128, 311
consolidation, 9
Consortium for Energy Efficiency (CEE), 42
container-based data center, 154
continuous data protection (CDP), 78
converged enhanced Ethernet (CEE), 263, 293
Converged Networks, 280
Cooling and HVAC, 138
cooling capacity, 30, 39
Cooling Challenges, 30
cooling requirements, 306
cores, 175
cost-per-capacity, 331
cost-per-gigabyte, 59
CPU processor utilization, 183
CRAC, 139, 158, 253, 292
CRAC plenums, 140

D

Data and Storage Security, 228
Data Archiving, 313
Data Center Challenges, 124
Data Center Container, 153
data center design, 129
Data Center Electrical Power, 132
Data Center I/O, 250

ta Center Location, 151
ta Center Tips, 158
ta Centers, 15
ta Compression, 231, 313
ta de-duplication, 314
ta Footprint
  Best Practices, 205, 208, 229
  Reduction, 7, 229, 247
  Techniques, 229
ta Life Cycle, 210, 297
ta Life-Cycle Patterns, 317
ta Preservation, 82
ta Proliferation, 207
ta Protection, 70, 71, 74
ta Protection Management, 85
a protection techniques, 86
ta Security, 69, 307
ta Storage, 205, 207
  Challenges, 206
  Issues, 206
  Trends, 206
tabase of State Incentives for Renew-
  bles & Efficiency, 42
 Power, 136
)R/RAM, 219
-duplication, 230, 232
mand Drivers, 167, 251
nse Wave Division Multiplexing
  DWDM), 260
partment of Energy, 5
ICP, 268
ital video recorders (DVRs), 16
MM, 184
ct attached storage (DAS), 213
ct current (DC), 133
ster recovery, 152
ster recovery (BC/DR), 251
ster recovery (DR), 56
k-to-disk (D2D), 73, 223
S, 269
ninant fuel source, 11
 pipe sprinkler, 146

DSIRE, 42
DWDM, 261, 273
dynamic data growth, 41
dynamic RAM (DRAM), 179

E

eBooks, 211
ECCJ, 44
ECKD, 260
Ecologically Friendly Data Centers, 29
eDiscovery, 87
Effective Data Center, 123
EHS, 4, 21, 150
EHS legislation, 17
electric utility, 133
ELectrical Power, 33, 133
Electrical power cables, 133
Electrical Power Grid, 35
electronic magnetic interference (EMI),
  153
ELV, 44
Emerging Technologies, 324
emission tax schemes (ETS), 12
Emulation, 188
Enable High Availability, 302
enabling enterprise IT, 330
Enabling Transparency, 54
Energy Avoidance, 25, 39, 301, 318
Energy Consumption, 20, 39, 298
Energy costs, 4, 61
Energy Efficiency, 9, 52, 319
Energy Efficiency Incentives, 41
Energy Efficiency Scenarios, 25
energy efficient
  green, 123
Energy Management, 132
energy rebates, 329
Energy Star, 44, 178, 318, 321, 327
Energy Usage, 37, 246
energy-effective data storage, 209
energy-efficiency incentives programs, 15
energy-efficiency perspective, 309

energy-efficient, 331
energy-efficient servers, 7
energy-per-gigabyte, 59
Enhanced Ethernet, 284
Environmental Health, 150
Environmental Health and Safety, 129, 320
environmental health and safety (EHS), 323
Environmental Health and Safety Standards, 45
environmental impact, 206
Environmental Protection Agency (EPA), 31, 178
environmentally safe cabling, 292
EPA Power Portal, 43
EPEAT, 44
ESCON, 214, 260
Ethernet, 260, 261
ETS, 15
E-Waste, 320
Exhaust Ducting, 140
Expanding Data Footprint, 207
external storage systems, 9

F

fabric-based appliances, 325
Facebook, 318
Facilities Review, 320
Fat-Capacity HDDs, 221
FCIP, 268
FCoE, 198, 282
fees for floor space, 208
Fibre, 316
Fibre Channel, 225, 262, 264, 280, 316
Fibre Channel deployments, 262
Fibre Channel HDDs, 214
Fibre Channel over Ethernet (FCoE), 262
Fibre Channel SAN, 311
Fibre over Ethernet (FCoE), 262

FICON, 214, 260, 262, 263, 272, 28 293
FICON/Fibre Channel adapter, 181
Financial incentives, 300, 329
Fire Detection, 144
fire suppression, 146
fire triangle, 144
FLASH, 219
FLASH SSD, 267
FLASH-based solid-state disk (SSD), 173
FM200, 146, 148
FORTRAN, 168
frozen capital, 126
Fuel cells, 137
fuel costs, 31
fuel sources, 11
fuel-efficient aircraft, 13

G

G&T, 3, 5
G&T Distribution, 36
gasoline, 11
General Data Retention, 230
Generalized Computer Hardware Arc tecture, 172
generating and transmission (G&T), :
Generator, 12, 135
geothermal, 43, 142
gigabytes (GB), 16
global carbon dioxide, 8
Global Positioning System (GPS), 22
Gordon Moore, 164
graphical user interfaces (GUIs), 167
Green and PCFE issues, 8
green and virtual data, 321
green and virtual data center, 324
Green Areas of Opportunity, 300
green data center, 52
Green Gap, 15
green gap, 5, 30, 323
green IT, 7

een PCFE Strategy, 22
een Spotlight, 20
een supply chains, 16
een Virtual Data Center, 297
een washing, 13, 332
eenhouse gasses, 10
d, 201
owing Green Gap, 5

H

bitat for Technology, 60, 123, 134
lon, 148
on, 146, 329
d disk drives (HDDs), 173, 265
rdware costs, 9
DDs, 208
at pumps, 142
h availability (HA), 72, 76
h-capacity JBOD, 318
h-capacity parallel ATA (PATA), 267
lographic Storage, 223
st bus adapter (HBA), 173
st bus adapters (HBAs), 253
TP, 268
/AC, 31, 34, 134, 139, 253, 292
brid Data Center, 142
brid Data Footprint Reduction, 234
dro, 43
drogen, 137

I

) connections, 292
) Connectivity, 181, 291
) networking, 251
) optimization, 175
) performance Gap, 58
) Protocols, 259
) security, 291
) Virtualization, 275
, 164
M, 211
plement Metrics, 308
entives and Rebates, 301

increased data footprint, 208
InfiniBand, 280
InfiniBand (IBA), 264
InfiniBand (SRP), 282
InfiniBand IOV, 286
InfiniBand SAN, 198
information life-cycle management
    (ILM), 211, 317
information services, 15
Infrastructure resource management
    (IRM), 66
infrastructure resource management
    (IRM), 59, 65, 123, 234, 325
inline memory modules (DIMMs), 179
input/output (I/O), 167, 249
input/output (I/O) characteristics, 207
Instant Messages (IM), 15, 251
instruction set architecture (ISA), 176
integrated circuit (IC), 164
Intel SpeedStep, 170
Intel VT, 176
intelligent fiber optic management, 292
Intelligent Power Management, 7, 224
Intelligent power management (IPM),
    224
intelligent power management (IPM),
    319
intermodal, 153
intermodal containers, 153
International System of Units (SI), 180
Internet Engineering Task Force (IETF),
    284
Internet Protocol (IP), 268
Internet-based data management, 252
IOmeter, 183
IOPS, 216
IOV, 275
IPM capabilities, 225
IRM, 59, 249, 298
IRM Activities, 67, 68
IRM Best Practices, 301
IRM Functions, 67

iSCSI, 198, 262, 263, 268, 282
IT, 3
IT Data Center Cooling, 140
IT data centers, 27
IT environments, 19
IT equipment, 60, 321
    special purpose, 145
IT Equipment Density, 127
IT factory, 50
IT Infrastructure, 65
IT PCFE, 30
IT professionals, 132
IT Resource Computing, 57
IT Resource Inefficiency, 299
IT resources, 54, 298
IT services, 327

J

Java Runtime Environment (JRE), 194
Java virtual machine (JVM), 194
JBOD, 266
JEDEC, 44
JEITA, 44
J-MOSS, 44
Joint Electronic Device Engineering
    Council, 44
JPEG, 206

K

Keyboard-video monitor-mouse (KVM),
    182
kilovolt amperes (kVa), 133
kilovolts AC (kVA), 133
kilowatt-hours, 38
kilowatt-hours (kWh), 133
KVa, 133
KVM, 184
kWh, 37

L

L1, 179
L2, 179
LAN-based backup, 78

large-capacity SATA, 316
LEED, 3, 44
legacy equipment, 9
line interactive UPS, 135
liquefied propane, 43
Liquid cooling, 170, 177
liquid pumps, 170
Local Area Networking, 258
logical unit number (LUN), 68, 229
LPG, 12
LUN, 68, 243

M

magnetic disk drives, 313
Magnetic Tape, 223, 313
MAID, 40, 246, 319
MAID 2.0, 224, 225, 319
MAID Level 3, 225
MAID systems, 225
mainframes, 174
managed service provider (MSP), 83
Managed Services, 155
Management Insight, 59
Mask-or-Move Issues, 309
maximize cooling efficiency, 178
Media Access Control, 282
Memory, 178
Memory and Storage Pyramid, 180
memory mirroring, 179
Metro Ethernet Microwave, 269
mezzanine (daughter) cards, 187
Microsoft ESRP, 183
Microsoft Exchange Personal, 215
Migration, 75
Misdirected Messaging, 5
Montreal Protocol, 146
Moore's Law, 164, 165
MP3, 206
MP4, 206
MPLS, 272
MSDS, 44

N

Port ID Virtualization (NPIV), 198
FS3, 145, 146
S, 186
ural gas, 11, 12
BS, 136
work attached storage (NAS), 83
twork File System (NFS), 157, 254
work interface cards (NICs), 192, 275
twork operations control centers
NOCCs), 253
twork Resource Management, 86
work-attached storage (NAS), 175
tworking Adapters, 311
tworking Demands And Challenges,
50
working interface cards (NICs), 173
tworking Options, 288
tworking Servers and Storage, 249
v generation data centers, 153
v habitat, 10
xt-Generation Data Center, 49
S, 269
IV, 277, 278
DC, 44
lear, 43
RAM, 178

O

line UPS, 135
crisis, 7
aro, 86
ine transaction processing (OLTP),
31
rating budgets, 126
erating System Containers, 194
rations per second (IOPs), 211
portunities for Action, 5
ical drives, 313
tical networks, 273
cle with NFS, 215
haned Resources, 88

orphaned storage, 89
Out-Sourced, 155
Overconsolidating, 193
Overcooling facilities, 125
Overhead Ceiling, 141
Oversubscription, 175, 289

P

PACE, 66, 226
PACE Impacts, 237
Pacific Gas and Electric (PG&E), 42
Parallel NFS (NAS), 269
Parallel NFS (pNFS), 254
paravirtualization, 192
PCFE, 3, 6, 14, 30, 34, 45, 66, 82, 90,
    160, 166, 202, 206, 210, 211, 212,
    249, 293, 304, 323
PCFE and Green Issues, 300
PCFE compliance, 17
PCFE constraints, 58
PCFE Consumption, 33
PCFE footprints, 16
PCFE Issues, 170, 226
PCFE resources, 310
PCFE Storage Issues, 209
PCFE Trends, 10
PCFE-related issues, 160
PCI SIG IOV, 283
PCI Single-Root Configuration, 257
PCIe, 184, 257
PCI-SIG IOV, 282
PCIx, 184
PDAs, 70
peak demand, 38
Performance Planning, 89
performance, availability, capacity, and
    energy (PACE), 247
Peripheral Component Interconnect
    (PCI), 181, 256
Peripheral Devices, 181
Perl, 168

personal digital assistants (PDAs), 16, 206
Phased cooling, 143
philanthropic motive, 14
Photoelectric devices, 145
physical IT resources, 124
physical machine (PM), 302
Physical Resources, 51, 56
Physical Security, 143
Physical Servers, 172
Physical structural considerations, 143
Physical-to-virtual (P2V), 199
PKZIP, 231
Plenums, 140
Point-in-time (PIT), 72, 77
Policy management, 87
Power consumption, 126
Power distribution units (PDUs), 133
Power Supplies, 182
Powering a Data Center, 134
precision cooling, 328
program counter (PC), 176
program status words (PSWs), 176
Provisioning, 188
Proxy-Based Backup, 79
Public consumer, 10
Public Law 109-431, 5

Q

quality of service (QoS), 315

R

radio frequency ID (RFID), 16
radio frequency identification (RFID), 70
radio frequency identification device (RFID), 229
radioactive americium-241, 145
radiofrequency interference (RFI), 153
RAID, 68, 83
RAID Affect, 236
RAID level, 239

RAID-enabled disk storage, 76
raised floor tiles, 159
RAM, 178
random-access memory (RAM), 173
raw device mapping (RDM), 198
REACH, 44
Read-only memory (ROM), 173, 179
Rebates, 41
recovery point objectives (RPO), 74
recovery time objectives (RTOs), 74
Recycling, 10, 21
Reduced Data Footprint, 312
redundant array of inexpensive disks (RAID), 68, 173
Relative Footprints, 128
reliable energy, 10
Remote Data Protection, 84
Remote Data Replication, 81
remote office and branch offices (ROBO), 83, 252
Removable hard disk drives (RHDDs), 173
removal of hazardous substances (RoHS), 22, 324
Resource Management, 65
RFID, 70, 150
RoHS, 44
ROMB, 173
RPC, 269
RPO, 77, 306
RTO, 77, 306
RTSP, 269
Ruby on Rails, 168

S

SaaS, 310, 313
safe technology disposition, 10
Safety Management, 150
SAMBA, 254
SAS storage technology, 266
SAS/SATA, 184
SATA, 314

SI Remote Protocol (SRP), 264
ure socket layer, 157
:uring IT Resources, 307
:urity, 40, 290
gmentation, 188
niconductor disk drives (SSDs), 218
ial ATA (SATA), 259, 265, 267
lal Attached SCSI (SAS), 265
ver and Storage, 58
ver Challenges, 164
ver Compute Capabilities, 167
ver Footprint Evolution, 171
ver Issues, 164
ver Provisioning, 303
ver Virtual Convexity Features, 279
ver Virtual Machines, 197
ver Virtualization, 9, 62, 188, 196
vers Physical, 163
vers Virtual, 163
vice Characteristics, 227
vice-level agreements (SLAs), 24
vice-orientated architecture (SOA),
326
nple Network Management Protocol
(SNMP), 253
all office/home office (SOHO), 130
all to medium-size business (SMB),
212
Bs, 252
oke and fire, 144
oke and Fire Suppression, 138
apshots, 72, 77
MP, 269
tware as a service (SaaS), 326
twarc dcvclopmcnt kits (SDKs), 71
HO, 130
id-State Devices (SSDs), 219
id-state disks (SSDs), 210
NET, 272
ice-Saving Clones, 236
EC, 183
D devices, 326

SSL, 157
Standby Power, 134
storage area network (SAN), 83
storage as a service (SaaS), 83
Storage Energy Efficiency, 244
Storage Management, 62
storage resource management (SRM), 86
storage service providers (SSP), 326
Storage Virtualization, 240, 241
storage-related resources, 209
Structured cabling, 144
suspended ceiling, 140

T

TCP/IP, 268
Technology Domains, 67
Texas Memory Systems (TMS), 221
thermal assessment, 320
thermal management technology, 328
Thin Provision, 236
Tier 1 Data Center, 130
Tier 2 Data Center, 131
Tier 3 Data Center, 131
Tier 4 Data Center, 132
Tiered Access, 255
Tiered Data Centers, 130
Tiered Networks, 62
Tiered Servers, 61, 316
Tiered Software, 61
Tiered Storage, 62, 212, 226
Tiered Storage Architectures, 213
Tiered Storage Media, 218
Tiered Storage Options, 228
Tiers Of Servers, 183
time-division multiplexing (TDM), 263
tin-wrapped software, 175
TiVo, 16
TPC, 183
Transmission Control Protocol (TCP),
268
Transparent Management, 55

U

U.S. Public Law 109-431, 5
underutilized IT resources, 54
Underutilized Storage Capacity, 234
uninterruptible power supplies (UPS), 133, 168
uninterruptible power supply (UPS), 220
unstructured file data, 206
UPS, 3
USB, 70
USB 2.0, 181
USGBC, 44

V

Variable Energy, 337
VCB, 80
VCB Proxy-Based Backup, 80
vendors, 132
VESDA detection, 145
VESDSA, 145
VIO, 275
Virtual Data Center, 49, 50, 52, 152, 269
    Components, 56
Virtual Environments, 70, 160, 288
Virtual I/O, 62, 275
Virtual Iron, 61
virtual LAN (VLAN), 261
virtual machine (VM), 52, 175, 302
Virtual Machine Movement, 75
virtual machines, 71
virtual office, 8
Virtual Offices, 63
virtual private network (VPN), 276
virtual SAN (VSAN), 276
virtual SCSI disk, 198
Virtual Server Environments, 72
Virtual Servers, 187
Virtual tape libraries (VTLs), 62, 79, 215, 242
virtual tape systems (VTS), 62
Virtualization, 40, 46, 49, 189, 321

Virtualization and Storage Services, 24
virtualization technologies, 57, 64
Virtualized, 191
Virtualized abstraction, 152
Virtual-to-physical (V2P), 199
Virtual-to-virtual (V2V), 199
VM virtual disk (VMDK), 191
VMDK, 198
VMware, 61
VMware Consolidated Backup (VCB) 80
vSafe, 157

W

wave and tidal action, 43
Wave Division Multiplexing (WDM), 260
WDM, 261
Web 2.0, 186
Web-based SSL, 158
WebEx, 251
WEEE, 44
WGBC, 44
Wide Area Application Service (WAAS 274
Wide Area File Service (WAFS), 274
Wide Area Networks, 269
WiFi, 174
WiMax, 174
wind, 43
Windows Common Internet File Syste (CIFS), 157
World Wide Node Names (WWNN) 277
World Wide Port Names (WWPN), 2
WysDM, 86

Y

YouTube, 318

Printed and bound by CPI Group (UK) Ltd, Croydon, CR0 4YY

23/10/2024

01778263-0015